Transforming Gender, Sex, and Place

Transgender, gender variant and intersex people are in every sector of all societies, yet little is known about their relationship to place. Using a trans, feminist and queer geographical framework, this book invites readers to consider the complex relationship between transgender people, spaces and places.

This book addresses questions such as, how is place and space transformed by gender variant bodies, and vice versa? Where do some gender variant people feel in and / or out of place? What happens to space when binary gender is unravelled and subverted? Exploring the diverse politics of gender variant embodied experiences through interviews and community action, this book demonstrates that gendered bodies are constructed through different social, cultural and economic networks. Firsthand stories and international examples reveal how transgender people employ practices and strategies to both create and contest different places, such as: bodies; homes; bathrooms; activist spaces; workplaces; urban night spaces; nations and transnational borders.

Arguing that bodies, gender, sex and space are inextricably linked, this book brings together contemporary scholarly debates, original empirical material and popular culture to consider bodies and spaces that revolve around, and resist, binary gender. It will be a valuable resource in Geography, Gender and Sexuality studies.

Lynda Johnston is Deputy Dean, Associate Dean Academic and Professor of Geography at the University of Waikato. The current Chair of the International Geographical Union Commission on Gender and Geography (2016–2020), Lynda has also served as Editor for *Gender, Place and Culture: A Journal of Feminist Geography*. *Transforming Gender, Sex, and Place* builds on previously published research, such as: gender and body building; pride politics and festivals; activism; drag queens and kings; queer sporting events and urban tourism.

Gender, Space and Society

Series Editors:
Peter Hopkins, Newcastle University, UK
Rachel Pain, University of Durham, UK

The series on Gender, Space and Society publishes innovative feminist work that analyses men's and women's lives from a perspective that exposes and is committed to challenging social inequalities and injustices. The series reflects the ongoing significance and changing forms of gender, and of feminist ideas, in diverse social, geographical and political settings and adopts innovative methodological and philosophical approaches to understanding gender, space and society.

For more information about this series, please visit: www.routledge.com/Gender-Space-and-Society/book-series/ASHSER1355

Transforming Gender, Sex, and Place

Gender Variant Geographies

Lynda Johnston

Routledge
Taylor & Francis Group

LONDON AND NEW YORK

First published 2019
by Routledge
2 Park Square, Milton Park, Abingdon, Oxon OX14 4RN

and by Routledge
52 Vanderbilt Avenue, New York, NY 10017

First issued in paperback 2020

Routledge is an imprint of the Taylor & Francis Group, an informa business

British Library Cataloguing-in-Publication Data
A catalogue record for this book is available from the British Library

Library of Congress Cataloging-in-Publication Data
A catalog record for this book has been requested

ISBN 13: 978-0-367-58774-1 (pbk)
ISBN 13: 978-1-4724-5479-9 (hbk)

Typeset in Times New Roman
by codeMantra

For my secondary school geographer teacher, Colleen Dennison, for making a world of difference

Contents

Figures

Preface

'I have never really understood gender or sex, and I still don't really understand it' one of my research participants said as we neared the end of a long interview. We sat at the dining room table in a wooden cottage in a small rural town in Aotearoa[1] New Zealand. Yann[2] retold their life story to me, one that started as intersex, followed by baby and childhood surgeries to create a 'female' body, and later, aged 50, gender transitioning. Yann felt comfortable with male hormones: 'not that I wanted to be a man, but because I wanted to see whether I could undo any of the childhood surgeries'. Yann went on to say: 'I have been trans-ing for ten years, yet, I don't identify as trans. I am in an in-between place'.

I begin with some of Yann's story as it is evocative of the ways in which gender diversity is embodied, lived, contested and in-between. The more gender variant geography research I conduct, the more I too 'don't really understand gender'. This is perhaps a surprising confession to make at the start of a book *on* gender variant bodies, spaces and places. Yet, after decades of conducting research on space, place and gender – bodies that conform or resist male / female binaries – the more the category of gender unravels.

Transforming Gender, Sex, and Place draws on research – interviews with people who live beyond biological norms associated with binary gender, as well as media representations, reports, and events – in order to tell stories. Stories can be powerful. They are the interweaving of philosophical thought, cultural codes, worldviews, contemporary and historical knowledges, spaces, places and lived realities. The potential of stories is that they can be constructed in any form, in any context, and through a variety of media. This book, then, is a collection of stories, as they were told to me, some of my own, and stories I have found. I have cared for them, nurtured them, and drawn on wider historical, social, political and spatial contents to make connections across stories, across places and to power relations. In order to understand the complexities of gender variant people and places I take a queer embodied geographical approach, which pays attention to identities, subjectivities, bodily senses, moods, sensations and feelings of being in and / or out of place. In other words, this approach prioritises the lived and deeply felt experiences of gender variant people and places. I also draw heavily on transgender studies, trans, queer and feminist geographies, feminist and gender studies, and sociology. A multidisciplinary approach is needed to understand the work that the gendered binary – male / female – does to produce normative places and bodies, and ways in which people and places trouble this binary. Through my research and community activism I have come to understand gender variant geographies as reflective of both belonging and alienation in everyday lives. Transgender people's embodied experiences – as discussed in this book – illustrate the concurrence of belonging *and* estrangement. Exposure to discriminatory actions triggers, in some places, acute sensitivity and vulnerability of the self. Yet, there are also examples of claiming and making inclusive transgender places, in which ideas and dreams are pursued.

I am particularly cognisant of my own embodied perspective, socio-spatial and personal histories as this experience inevitably imprints itself onto the pages of this book. This approach, I hope, reflects the lived realities of participants, other scholars and their research, activists, media and myself. The stories are close to my heart and my own gendered experiences. For as long as I can remember, I have questioned all social conventions and spaces that appear to be naturally aligned with restrictive articulations of gender. I write this book as an academic, in a small city of approximately 160,000 people, called Hamilton, in the North Island of Aotearoa New Zealand. Yet, I also bring to the pages my experience of growing up in a small rural village called Waitati, in the lower South Island. I am grateful that my parents didn't apply 'gender rules' to their children, but I was certainly aware of what other children were supposed to do, and how they should appear, according to their given gender. Similar to Yann's rural home, I grew up in an out of the way place where people tended to embrace difference: different ways to live sustainable lives with low or no income; different ways to do relationships; and, different ways to do gender and sexuality. This had a profound impact on me and my sense of social justice. I quickly identified as a feminist as soon as I knew what the word meant and in retrospect, this was my first attempt to challenge gender binaries. I relished any activities and spaces that were not traditionally deemed to be for 'girls' in the late 1960s and 1970s. Now into my fifth decade I continue to take and create opportunities to queer gender, sex and place. In many places I feel a sense of unease between how my body appears and how people react to me, with regards to sex and gender. I experience marginalisation and exclusion due to the intersections of gender, sexuality and even body shape and size. I am Pākehā (a Māori term for white settler society New Zealanders) and aware of how privileged spaces open up for me. While I grew up in a household with low income, I am now a well-paid university professor. I bring these paradoxical and intersectional lived experiences to the pages of this book.

At the heart of *Transforming Gender, Sex, and Place* is a political imperative for human rights, civil rights, in short, rights to be whichever gender you are (including not having a gender). The word 'transforming' in the title of the book is, of course, a nod towards the term 'transgender', yet it also symbolises the political importance of troubling hegemonic sex and gender as they are mutually constructed through bodies, places and spaces. The book invites readers to think about sex and gender as always in the making in and through place, and to explore 'categorical crossings, leakages, and slips of all sorts, around and through the concept 'trans-' (Stryker et al. 2008, 11).

Transgender, gender variant and intersex people exist in all societies yet little is known about their relationship to place. Many gender variant people experience discrimination, oppression and marginalisation in relation to specific places and spaces. Some spaces and places create geographies of belonging for gender variant people. There has been limited research about gender variant people's feelings of 'being in and / or out of place'. As a consequence, cisgender people who conform to a male / female gender binary may have little to no understanding of the issues affecting gender variant people's lives.

Transforming Gender, Sex, and Place includes original research in the form of interviews, participatory methods and cultural texts. It includes interviews with 22 people who identify as transgender and gender variant (in various ways) as well as intersex. These interviews, which range in length from one to three hours, have been conducted in Aotearoa New Zealand (in cities and in rural places), as well as in some U.S cities. I also bring to the book (implicitly and explicitly) my previous research on: body builders (Johnston 1996); drag queens (Johnston 2012); Coccinelles (transwomen in Israel) (Misgav and Johnston 2014); drag kings (Johnston 2009); the world's first transgender mayor (Johnston and Longhurst 2010); small towns, conservative cities and 'coming out' as trans (Johnston and Longhurst 2013); and, intersex (Johnston 2014).

Throughout my research I remain guided by feminist and queer methodologies and the insistence to always be aware of researcher positionality. I practise self-reflexivity and prioritise 'non-hierarchical interactions, understanding, and mutual learning, where close attention is paid to how the research questions and methods of data collection may be embedded in unequal power relations between the researcher and research participants' (Sultana 2007, 375–376). For my research projects, practising self-reflexivity means recognising that I do not experience transphobia, sexism, patriarchy, classism, homophobia and racism in the same way as each research participant.

It is my hope that this book will help readers develop a deep appreciation of the diversity and complexity of gender variant embodied experiences of place, space and power. I show the ways in which bodies are gendered and sexed through different social, cultural and economic networks and through different spaces and places. I draw on a number of European and U.S. examples but I also include examples from elsewhere – Aotearoa New Zealand, Australia, South Pacific, South America, Israel, and parts of Asia – in order to convey to readers that place is integral to the production of genders.

Notes

1 Aotearoa is the Māori term for New Zealand. Te reo Māori (Māori language) is an official language of Aotearoa New Zealand since the passing of the 1987 Māori Language Act. Aotearoa is used frequently, both formally and informally. It is common for institutions, and indeed government departments and ministries, to have Māori names alongside English names. Throughout the book I use Aotearoa both with and without New Zealand to acknowledge the nation as bicultural and to continue to contest the ongoing colonisation. Place naming continues to be contested (see Berg and Kearns 1996, Larner and Spoonley 1995 for commentary on the colonial politics of place naming).

2 Many research participants have pseudonyms to protect their anonymity. Some insisted I use their actual names.

Acknowledgements

The contents of this book are drawn from two decades of research on gender variant people and places. I am indebted to my family, friends and colleagues who have sustained me in many ways and for many years: emotionally; intellectually; and, physically.

I am fortunate to be part of communities, who, with their support, make my life liveable and this book possible. To the many people who agreed to be part of my research projects over the last two decades, I thank you. There are numerous acts of support, all of which are vital to this project. I deeply value the mundane acts of sharing of conversation over coffee or dinners, as well as the extraordinary events, such as celebrations of gender diversity and memorial services. I owe a tremendous debt of gratitude to our gender variant people in Hamilton, where I live, but also across Aotearoa. We are a small country, both in land and population size. The beauty of being in a small country, and at the 'bottom of the world', is that it becomes easier to connect to others who embody genders beyond a binary model. I want to acknowledge the group Hamilton Pride Inc., in particular, for being 'a base' for me for many years. I also wish to acknowledge the group Te Rākei Whakaehu, also based in Waikato, for their enduring commitment to healthy lives for those who identify as transgender and Māori, and from whom I continue to learn a great deal.

My colleagues at the University of Waikato play a significant part in supporting the ideas contained in, and the production of, this book. Professor Robyn Longhurst continues to inspire and support my research and life in most ways! Robyn is always willing to discuss ideas, listen to my frustrations, and offers more than one could ask for in intellectual companionship and friendship. It is because of Robyn, more than two decades ago, that I was brave enough to become a researcher on topics that are usually marginalised. Dr Gail Adams-Hutcheson has been a wonderful research assistant for many years. I'm grateful for her critical insights, collegiality and cheerful transcribing of interviews. Gail has also played a vital role with editing and organising materials for this book. Sandi Ringham also worked as a research assistant, helping with interview transcription and discussing ideas contained in this book. Associate Professor John Campbell shared with me many news items and stories from the Pacific, particularly about fa'afāfine in Samoa, and Dr Colin McLeay regularly found media items that have been useful in this book.

I'm fortunate to be in a faculty that has a growing number of scholars conducting research on transgender issues. The research by Dr Johanna Schmidt, Professor Katrina Roen and Dr Jaimie Veale is inspirational. In 2016, and in collaboration with Jaimie, Renée Boyer, and a handful of students, we started a Rainbow Alliance group on campus at the University of Waikato. It's been a small yet important way to challenge cisgender and heteronormative spaces on campus. Thank you to Jaimie and all those who support this campus initiative.

While writing this book I have been inspired by my PhD candidates – Tegan Baker, Sunita Basnet, Elaine Bliss, Lisa Melville, Grace O'Leary, Sandi Ringham, Anoosh Soltani, Kathy Ullal – and I thank them for their critical scholarship, which challenges social and spatial injustices. I am also grateful for the warm collegiality of Associate Professor Lex Chalmers, Rachel Gosnell-Maddock; Brenda Hall, Betty-Ann Kamp, Kate Mackness, Paula Maynard, Max Oulton, Dr Naomi Simmonds, Dr Pip Wallace and Professor Iain White. Heather Morrell, a geographer and university librarian, assisted with references and purchased books I requested for the library. Heather's keen interest in research is always appreciated. The Faculty of Arts and Social Sciences have always supported my applications for research grants, which helped me juggle research, teaching and administrative tasks for the university. During the last two years I took on senior leadership roles in my faculty and I am grateful to my Dean – Professor Allison Kirkman – for her support, which enabled me to meet my various administration and research commitments. Up until her retirement at the end of 2017, I worked closely with Carin Burke, our faculty's Academic Services Manager. Carin kept all systems running smoothly, which meant I could have time to focus on this book.

I would also like to acknowledge the global network of scholars whose work inspires me. The *Gender, Place and Culture* editorial team and community of scholars have a profound impact on my research into bodies, genders and sexes. Special thanks to Professor Petra Doan – a member of the *GPC* editorial board – whose excellent research spans geography, planning, transgender and feminist studies. The International Geographical Union's Geography and Gender Commission members are an impressive group of people and their commitment to difference and diversity is uplifting.

Routledge editorial assistants Ruth Anderson and Priscilla Corbett, plus editor Faye Leerink provided wonderful guidance and, at times, reassurance. They reminded me of the deadlines, and helped ensure the smooth production of the book. Ting Baker provided exemplary copyediting during the final stages of book production.

Some of the material in the book was presented at conferences. I am grateful for audience members and their engagement with the ideas I presented. These discussions have undoubtedly informed and transformed my thinking. Some of the information in Chapter 1 on transgender masculinities was presented at the Association of American Geographers' 108th Annual Meeting, 29 March – 2 April, San Francisco. Chapter 1 also contains material that was presented at the Joint Institute of Australian Geographers and New Zealand Geographical Society Conference, 30 June – 2 July 2014, Melbourne. Parts of Chapter 4 on bathrooms was presented at the New Zealand Geographical Society Conference, 'Interventions' Otago University, 1–4 February 2016, Dunedin. Some of the material in Chapter 6 on workspaces was presented at the 'Gendered Rights to the City: Intersections of Identity and Power, Conference', co-organized by the International Geographical Union Commission on Gender and Geography, the Geographical Perspectives on Women Specialty Group of the Association of American Geographers, and the Department of Geography, University of Wisconsin, Milwaukee, 19–20 April. At the Association of American Geographers' 107th Annual Meeting, 21–25 April, Chicago, I presented another paper on workspaces and transgender embodiment. Chapter 9 contains information about intersex health and well-being, which was presented at the Association of American Geographers' 106th Annual Meeting, 8–12 April, Tampa, Florida.

My deep gratitude goes to my super supportive family – to Karen for always going above and beyond the role of sister (she even transcribed interviews for me!), and to our dad Murray who – years ago – refused to let prevailing gender norms restrict our childhood and is always interested in, and part of, our adult lives. Our unconventional mother – June – taught us to be tenacious and positive when faced with personal challenges, and while she passed away in 2005, remains a strong influence in all that I do. Finally, a huge thank you to my amazing partner, Tara, who despite her own demanding career and family commitments is unfailingly supportive of me.

1 Transgressive bodies and places
Bending binaries

> It's not all about that private 20-week ultrasound anymore! More and more parents are planning gender reveal parties to find out their baby's sex and share the excitement with family and friends through a big reveal, usually in cake form. So how is it done? Most moms and dads have their ultrasound technician write the child's sex on a slip of paper that is placed in an envelope. It's then dropped off at a bakery where a cake is baked in the appropriate color and iced to hide the news. When the parents-to-be cut into the sweet treat, they learn about their future offspring's sex.
>
> (Gruber 2016, no page number)

I begin with the above excerpt from the popsugar.com website, not to celebrate another commercially driven baby party opportunity, but to highlight the persistent thinking that bodies are born, and remain, either male or female. Gender-reveal parties are trending in western countries where most pregnant women undergo prenatal sonography, of which approximately 50 per cent want to know if they are having a boy or a girl (Shipp et al. 2004). The party centrepiece – a cake – conceals the secret below multicoloured icing. The first slice of the cake reveals sponge (or more icing) that is either pink – to indicate the impending arrival of a girl – or blue for a boy. The many possibilities of sex and gender are absent from these parties as they only take into consideration gender and sex as a binary. Parents-to-be, family and friends at gender-reveal parties all assume that the baby – regardless of its biology – will adopt a gender identity that is socially and culturally prescribed according to normative understandings of sexed bodies. Gender-reveal parties reinforce the notion that gender is synonymous with genitalia, and they assume a narrow understanding of sex and gender, despite:

> evolving notions about what it means to be a woman or a man and the meanings of transgender, cisgender, gender non-conforming, genderqueer, agender, or any of the more than 50 terms Facebook offers users for their profiles. At the same time, scientists are uncovering new

complexities in the biological understanding of sex. Many of us learned a high school biology that sex chromosomes determine a baby's sex, full stop: XX means it's a girl; XY means it's a boy. But on occasion, XX and XY don't tell the whole story.

(Henig 2017, 51)

Some people may wish to live gender free. Consider, for example, the decisions of New South Wales (NSW), Australian Registry of Births, Deaths and Marriages. In March 2010 they sent Scottish-born Norrie May-Welby an immigration certificate that states 'sex not specified'. The *Sydney Morning Herald* (Gibson 2010, no page number), in an article entitled 'Sexless in the city: a gender revolution' wrote:

> This Mardi Gras, Norrie received a gift that no other androgynous person in NSW has had before. The night before the parade, the postman [sic] brought a certificate from the Registry of Births, Deaths and Marriages that contained neither the dreaded 'M' nor its equally despised cousin, 'F'. Instead, it said 'sex not specified', making the 48-year-old Sydneysider, who identifies as neuter and uses only a first name, the first in the state to be neither man nor woman in the eyes of the NSW government. Because Norrie was born in Scotland (and used the surname May-Welby), it was not a birth certificate but a Recognised Details Certificate – the version given to immigrants who have changed sex and want it recorded.

The decision to allow Norrie May-Welby to immigrate to Australia as 'sex not specified' started a legal battle. The Registry backtracked, reversing their decision and claiming they did not have the legal authority to produce a gender neutral certificate. Norrie filed a complaint with the Australian Human Rights Commission and the Court of the Appeal. The legal battle went on for four years, resulting in a High Court ruling that the law does recognise a person may be neither male nor female. Norrie is delighted, claiming a victory for 'sex and gender diverse people throughout Australia' (ABC News 2014, no page number).

A quick glance at other global media tells us more about sex and gender. Media are reporting that a 'gender revolution' is taking place in and through screens and in particular places (see *National Geographic's* (2017) special issue on 'The Shifting Landscape of Gender'). Early in 2014, for example, Facebook announced a new list of gender identities, including 51 possible options, allowing users to select transgender, intersex, genderqueer and other possibilities. These options recognise gender identities as fluid, changing and complex (Baldwin 2014). Transgender and gender variant representation on television is hitting new highs with, for example, *I Am Cait* (2015) featuring Caitlin Jenner – formerly known as Bruce Jenner and a high profile Olympic athlete. *I Am Cait* is a reality television show that documents

Caitlin's gender transition. *Orange is the New Black* (2013) may be the first women's prison drama to include a transgender woman character played by an African American trans actress Laverne Cox. *Transparent* (2014) centres on the main character – Maura Pfefferman – and her gender transition and includes transgender actors in a variety of transgender and gender variant roles. The teen show *Glee* (2009) includes a variety of feminine men, masculine women, as well as sexually diverse characters.

From gender-reveal parties to television series that centre on transgender characters, sex and gender are central to bodies, places and spaces. For decades critical scholars – feminists and others – have argued that bodies don't simply exist, rather they are gendered, sexed, racialised (Davis 1997; Gatens 1988; Rose 1993) and that lived experiences associated with age, size, and class matter (Duncan 1996; Johnston 1997; Longhurst 2014; McDowell and Court 1994; Moss and Dyck 2002). Bodies and space intermingle (Probyn 2003) and may prompt 'gender trouble' (Butler 1990). One cannot escape the requirement to 'check' either a 'Male' or 'Female' box when applying for jobs, bank loans, enrolling at school, to name just a few institutional examples where place is imagined as only via a binary gender. Binding legal gender status appears on birth certificates, passports and drivers licences. While across the globe one can see the changing politics of naming race and ethnicity via official census questions (Yanow 2015), the same has not happened to sex and gender. This is because a narrow understanding of biology is commonly understood to dictate a male / female division. The notion, therefore that all bodies fit into either male or female is hotly contested and has been for some time. In a groundbreaking collection on geographies of sex, gender and sexuality, Julia Cream (1995, 35) asks: 'What is the sexed body?' Cream (1995, 36) shows that bodies that do not fit a male or female binary – bodies such as those who are transsexual, intersex, have XXY chromosomes – are 'pioneers placed, often unwillingly, at the frontiers of sex and gender' (see also Fausto-Sterling 2000; Kessler 1998; Preves 2003). During the mid-1990s Namaste (1996) addressed violence against people whose gender is understood as non-normative, particularly for gay men, lesbians and transgender people. By the end of the 1990s it became well established that bodies and places are mutually constitutive and performative (Nast and Pile 1998), are an important site for understanding power, and that gendered subjectivities are fluid.

In the U.S. transgender people report experiencing employment discrimination (see Chapter 6, and Grant et al. 2011; Lombardi et al. 2002). The 2011 National Transgender Discrimination Survey (Grant et al. 2011) shows that transgender people in the U.S. are far more likely than cisgender people to live in poverty, be homeless and unemployed. Transgender people are disproportionately represented in HIV statistics and in poor mental health. These disparities increase for transgender people of colour. Transgender people 'experience vulnerability at the hands of the state and capitalism; this vulnerability is compounded at intersections of marginalisation' (Herman 2015, 90).

This book argues that bodies, gender, and space are inextricably linked. *Transforming Gender, Sex and Place* brings together contemporary scholarly debates, original empirical material, and popular culture to consider bodies and spaces that revolve around, and resist, binary gender. While binaries have, for some time, been the subject of critique by human geographers (Cloke and Johnston 2005), they are being dismantled by geographers interested in trans theories and queer geographies (see Browne 2006; Browne and Lim 2010; Doan 2007, 2009, 2010, 2016; Hines 2007, 2010, 2013; Hines and Sanger, 2010; Hines et al. 2010; Hines and Taylor 2012; Johnston 2005a; Johnston and Longhurst 2016; Nash 2010a, 2010b, 2011). The overriding argument is that gender is fluid and dependent on place and space. Petra Doan (2010, 635) reminds us all that 'transgendered and gender variant people experience the gendered division of space as a special kind of tyranny – the tyranny of gender – that arises when people dare to challenge the hegemonic expectations for appropriately gendered behaviour in western society'. This book examines this tyranny in relation to place and space. I bring to the pages an embodied geography of transphobia and gender binarism. While it is clear that transphobia exists, it is far less evident what transphobia *does* to people and places. Talia Mae Bettcher (2014, 249) refuses to give a neat definition of transphobia, rather, saying that transphobia is:

> directed toward trans people. In doing this, I have tried to avoid smuggling an actual account of the underlying nature of transphobia into the definition. But much depends upon how the expression trans people is itself defined. If it is defined as 'those who violate gender norms,' or as 'those who are problematically positioned with respect to the gender binary,' then a very general account of the nature of transphobia is immediately forthcoming—namely, transphobia is a hostile response to perceived violations of gender norms and/or to challenges to the gender binary.

I focus on how transphobia is lived, experiences and embodied in the context of the everyday in particular spaces and places, and within interpersonal relationships. The book also celebrates gender diverse bodies, spaces and places. The following questions are addressed: how are place and space transformed by gender variant bodies, and vice versa? Where do some gender variant people feel in and / or out of place? What happens to place and space when binary gender is unravelled and subverted?

Transforming gender and sex

The book's title reflects a hopeful geography (Lawson 2005) informed by feminist, queer, and trans theories that hold firmly to the notion of materiality (in other words, I am not just interested in representations or social constructions of bodies but also in 'real' fleshy bodies). In order to understand how gender variant bodies transform place and space, I turn to queer, feminist and trans theories. Queer theory emerged in the past two decades primarily as a critique of heteronormativity (Warner 1991), and the assumption that bodies are easily divided into 'man' and 'woman', and that these categories are opposing, natural, and biological. Queer theory allows for the challenging of categories gender and sex, including the idea that 'sex' refers only to biological or scientific truth, while 'gender' – deemed its binary opposite – is socially constructed. Gender and sex intermingle and the matrix of heteronormativity may be disrupted when the biological diversity of bodies is recognised (Butler 1990). The existence and recognition of socially and culturally marginalised gender identities – transgender, transsexual, intersex, genderqueer, cross-dresser, masculine women, feminine men, and so on – also destabilises heteronormativity.

A defining aspect of trans scholarship is that it developed (mostly) independently from Women's Studies programmes, feminist studies, and LGBQ studies, despite many overlapping areas of interest. Susan Stryker (2006, 7) notes that trans scholarship has a 'vexed' relationship with both feminist and queer theory due, in part, to some feminists' attitudes towards trans-identified people. A popular and often derogatory expression, TERFs (transgender exclusionary radical feminists), is used to identify anti-trans groups (see Hines 2017; Williams 2016). Certainly, Janice Raymond's 1974 book *The Transsexual Empire* illustrates the overt hostility from feminists of this time. This hostile relationship – generated by Raymond's book and more recently by Sheila Jeffreys' 2014 book *Gender Hurts: A Feminist Analysis of the Politics of Transgenderism* – create further rifts between trans and feminist scholarship. Concerned that Jeffreys' book would add to anti-transgender sentiments, the *Transgender Studies Quarterly* journal created a special issue on Trans/Feminisms (Stryker and Bettcher 2016). Following this special issue, I too want to 'expand the discussion beyond the familiar and overly simplistic dichotomy often drawn between an exclusionary transphobic feminism and an inclusive trans-affirming feminism' (Stryker and Bettcher 2016, 5). I use queer and feminist theories that are trans inclusive in order to affirm gender diversity and experiences. Feminist transphobia is absolutely not a universally held belief and feminist politics are useful for understanding transgender lives. I take an inclusive view, drawing from various trans, feminist, queer and geographical studies, to argue for the recognition of multiple subjectivities, behaviours, politics, and lived experiences.

Paisley Currah, Jamison Green and Susan Stryker (2009, 3), leaders in the field of trans studies, provide critical definitions which this book builds on. They note the term 'trans' is:

> a sense of persistent identification with, and expression of, gender-coded behaviors not typically associated with one's sex at birth, and which were reducible neither to erotic gratification, nor psychopathological paraphilia, nor physiological disorder or malady. The self-applied term was

meant to convey the sense that one could live non-pathologically in a social gender not typically associated with one's biological sex, as well as the sense that a single individual should be free to combine elements of different gender styles and presentations, or different sex/gender combinations. At one level, the emergence of the 'transgender' category represented a hair-splitting new addition to the panoply of available minority identity labels; at another level, however, it represented a resistance to medicalisation, to pathologisation, and to the many mechanisms whereby the administrative state and its associated medico-legal-psychiatric institutions sought to contain and delimit the socially disruptive potentials of sex/gender non-normativity. Having an intelligible social identity is the means by which an individual body enters into a productive relationship with social power. Thus 'identity politics,' the struggle to articulate new categories of socially viable personhood, remains central to the consideration of individual rights in the United States, and to the pursuit of a more just social order. The emergence of 'transgender' falls squarely into the identity politics tradition.

Gender categories, then, cross and slip in and around the concepts of 'trans' (Stryker et al. 2008): trans-gender; trans-sex; trans-place; trans-space; and I encapsulate these terms within the broad field of gender variant geographies. Gender is the primary category of analysis, and the book advances the ways in which gender variance intersects with sex, place, and space. Transgender, as a term, has become a shorthand concept for multiple, sometimes overlapping, and over contested meanings. It may signal a type of gender crossing, yet it also may signal the multiple ways of embodying genders that muddle or queer the gender binary. The term is useful when advocating for identity-based rights claims and critical explorations of gender based inequalities. There are, however, ever evolving terms to describe embodied identities that are gender nonconforming and that resist the coercive binary system. In this book I use several terms that are used by participants, community and activist groups, plus intersex and transgender scholars. I use the words 'trans', 'transgender', 'gender identity', 'gender expression', 'intersex' and 'gender variant' frequently. I use 'trans' as an inclusive and far reaching term to indicate people who identify as transgender, transsexual, or within the transgender spectrum. The term 'transsexual' usually refers to a medical expectation or medical transition that may be required by doctors when evaluating who is eligible for gender reassignment procedures. All of these terms are political, and indicate a wide variety of people whose gender identity or expression transgresses binary gender norms. I also use the word 'gender variant' because there are many people who may not identify as 'trans' or 'transgender' yet who experience persistent gender identity discrimination. Masculine women and feminine men, for example, though not trans-identified, experience gender identity discrimination. The term 'gender variant', hopefully, captures the way in which bodies and expressions of gender and sex 'push back' on the narrow understandings of 'maleness' and 'femaleness' according to the gender assigned to people at birth.

Other terms that are frequently used in this book are 'identities' and 'subjectivities'. Many commentators who address the intersectionality of gender, sex, race, class, ethnicity, sexuality and so on, focus on the pitfalls and possibilities of naming and analysing each category as not fixed but dynamic, changeable and interlinked (Hines 2007; Hopkins 2017; Lykke 2010; Probyn 2003; Rose 1995). Some scholars define identity and subjectivity against each other, so that identity tends to mean external social categories that individuals subscribe to, while subjectivity refers to the way people adopt social categories and turn them into lived choices and experiences (Longhurst 2003; Probyn, 2003; Wetherell, 2008). I keep these two concepts in tension throughout the book to show their mutual construction and the complexities of people's lived experiences.

There are, of course, many non-western gender expressions and identities, particularly in the South Pacific. It is impossible to list all South Pacific Indigenous gender identities and expressions and many of these are contested, depend on place, culture, and ongoing colonisation. In Aotearoa New Zealand 'takatāpui' is sometimes used as an umbrella term for gender and sexually diverse Māori,[1] the Indigenous peoples of Aotearoa. 'Takatāpui' first appeared in the 1871 *Dictionary of the Māori Language* (third edition) and means 'intimate companion of the same sex' (Williams 1971, 147). Rallying around the term takatāpui, prominent Māori scholar and activist Ngahuia Te Awekotuku (1991, 37) issued a challenge: 'we should reconstruct the tradition, reinterpret the oral history of this land so skilfully manipulated by the crusading heterosexism of the missionary ethic'.

Some Māori may use the term whakawahine, which literally means to be like a woman, or tangata ira tane, which means to be like a man. For Samoan people, the term fa'afāfine is often used to refer to being like a woman, while fa'atamaloa – or tomboys – is to be like a man. In Tonga, the term fakaleiti is to be like a woman, and in the Cook Islands akava'ine is to be like a woman, and in Fiji vakasalewalewa. These terms are heard in the Pacific region, yet, they are also contested, are performed differently depending on place, and, of course, there are many more gender identities and expressions (Besnier and Alexeyeff 2014). The inclusion of Indigenous people's lived experiences is vital if transgender knowledges are to be decolonised (Besnier 1997, 2002, 2004, 2011; Hutchings and Aspin 2007; Kerekere 2017; Schmidt 2010, 2016; Te Awekotuku 2001; Tcherkézoff 2014).

The term 'transgender' is usually understood – in the west – as an umbrella term for a diverse group of people whose gender is at odds with their ascribed gender (Valentine 2007). Trans* (with an asterisk) may be shorthand for numerous gender identities, for example, transgender, transsexual, transvestite, cross-dresser, trans women, trans feminine, trans man, trans masculine, genderqueer, non-binary, gender fluid, agender, pangender, non-gender, bi-gender, demi-gender, gender diverse, third gender, drag king and drag queen, to name just a few. Non-western non-heteronormative gender categories, for example those in the Pacific, do not necessarily align with these western understandings of transgender.

Often conflated with trans may be those who are intersex (Hird 2003). Intersex refers to a person whose body is not the 'standard' male or female type and it is estimated that one in 100 people have bodies that differ from standard male or female (see the Intersex Society of North America, www.isna.org). The term encompasses over 30 different conditions that all originate pre-natally during the foetus' development (Christmas 2010; Dreger 2004; Johnston 2014). The social privilege that comes with having normatively gendered bodies – where bodies feel comfortable enough to 'sink' into the space (Ahmed 2004) – is known as cisgender (Enke 2012). The terms 'cisgender' and its shorthand 'cis' are used to highlight the embodied privileges of non-transgender, non-gender variant bodies in particular places, such as bathrooms (Cavanagh 2010, and to unsettle the politics of naming women's research groups (Browne et al. 2013). The term 'cissexual' is usually defined as 'non-transsexual' and similarly, the term 'cisgender' tends to be defined as 'non-transgender'. Cis, cissexual, cisgender, cissexism, cisgender assumption, cis privilege and other related cis terms started to circulate in the early nineties and they hold great potential to understand embodied privilege (Cava 2016).

Multiple genders, and terms that trouble gender binaries, exist. For example, a person may identify as 'genderqueer' when they do not align with normative or conventional binary gender identities (Nestle et al. 2002). Building upon earlier concepts such as 'androgyny', transgender, feminist scholars and activists challenge and expand the concept of gender. Sandy Stone's *The Empire Strikes Back: A Posttranssexual Manifestation* (1987), and Leslie Feinberg's *Transgender Liberation* (1992) and *Stone Butch Blues* (1993) are key texts that help develop a genderqueer movement (Weaver 2014) and 'gender outlaws' (Bornstein 1994). Indeed, Stone (1987) and Kate Bornstein (1994) began queering transsexuality, giving rise to a growing number of transgender theories. The complexities and ambiguities of lived and embodied experiences are examined through a range of theories, as gender variant people conform to, or resist, existing gender expectations.

Key gender terms, then, rather than being fixed, are reflective of their geographical, historical and disciplinary context (Hines 2010; 2013). For example, the term 'transvestite' is often related to gender deconstruction, in other words, another type of gender logic (Garber 1992). Yet, this does not give the full lived experience of people who cross-dress and / or do drag performance. Care is needed to not undermine the position, status, and identity of gender categories such as transvestite (Namaste 1996). As noted, language is powerful and all categories are political and construct meanings. Gender variant and transgender terminology is highly contested and changes depending on time and place. Throughout the book I adopt some of the terminology that is used by activists, scholars and research participants.

Some people may use surgeries and / or hormones to align their bodies with a gender identity that differs from the sex they were assigned at birth. The term 'transsexual' was widely used for such individuals but is not as common today. It came from psychology scholars and practitioners in the U.S. during the 1940s (Meyerowitz 2002). It wasn't until the 1990s, and as a result of gender activists, that 'transgender' became an umbrella term for a wide variety of gender outlaws and gender nonconformers (Bornstein 1994; Valentine 2007). The term 'genderqueer' surfaced in the early 2000s as a result of people identifying with a blend of various aspects of femininities and masculinities (with or without bodily modifications) (Nestle et al. 2002). This term is a direct challenge to male / female – and associated binaries such as nature / nurture, mind / body and essentialism / social constructionism – that can be found in literature (Johnston 2005a, 2016) and some of the arguments are rehearsed here in this book. When discussing research projects I use terms that participants specifically asked me to use, for example, a participant in her 70s describes herself as transsexual, which is a term that had popular currency from the 1960s to the 1980s.

It may not surprise readers that there has been a rapid increase in the interdisciplinary field of gender variant and transgender studies. Gender scholars, geographers, sociologists, psychologists, and scholars in cultural and health studies have all shown a growing interest in gendered and sexed bodies that live and experience places beyond normative notions of male / female. In this work, gender is the central co-ordinate of embodied subjectivity. The relationship between bodies, space, place, gender and sex is explored in conference sessions, in articles in scholarly journals, some specifically about transgender (e.g. *TSQ: Transgender Studies Quarterly, International Journal of Transgenderism)* as well as other journals in cognate disciplines (e.g. *Sexualities, Gender, Place and Culture, Body & Society)*. Geographers have been troubling gender (and its relationship with sexuality) for some time in academic books (e.g. David Bell and Gill Valentine's (eds) (1995) *Mapping Desire*; Kath Browne, Jason Lim and Gavin Brown (eds) (2007) *Geographies of Sexualities*; Kath Browne and Leela Bakshi's (2013) *Ordinary in Brighton*; my book (2005b) *Queering Tourism: Paradoxical Performances at Gay Pride Parades* and another – with Robyn Longhurst – (2010) *Space, Place and Sex)*. In short, while the scholarly field of gender variant geographies is gaining a place in the academy, there is potential for deeper engagement within this field and beyond (Stryker and Currah 2014).

Transgender studies that focus on identities (Davy 2011; Ekins and King 1996; Ekins 1997; Prosser 1998) lay important foundations for understanding people's relation to, for example, medical care, services and prevailing discourses. Psychological research is an area that has been at the forefront of understanding transgender identities (Bockting and Coleman 1992; Chrisler and McCreary 2010). Notably, this scholarship has moved away from its foundations of 'abnormal psychology' (which imagined transgender expressions as a sign of mental illness or 'disorder'). Contemporary psychology

scholarship tends to focus on the individual and self-identity, and more research is required to understand the relational, contextual, and spaces associated with gender variance. Sociological trans scholarship has made great
advances in highlighting power, politics, subjectivities, and relationships
(see Hines 2013; Hines and Sanger 2010; Stryker and Aizura 2013; Stryker
and Whittle 2006), although this research tends to gloss over the nuances of
place and space. Furthermore, much of this scholarship is firmly located in
the U.K. and U.S.

Transgender scholars (Davy 2011; Prosser 1998; Stryker and Aizura
2013) have been deeply critical of the lack of engagement with transgender
lives. Much of this scholarship argues for not universalising transgender
subjectivities. The specificities of transgender mobility (Halberstam 2005),
migration (Cotton 2012), and trans movements for social change (Wilson
2002) are now on the agenda. Currah et al. (2009 3), leaders in the field of
trans studies, provide critical definitions, as well as trouble the term 'trans'.
The work of these scholars is slowly being incorporated in geographies of
genders and sexualities, as well as other disciplines. There is a great deal of
scope, however, to consider the lived and everyday realities of gender variant bodies at a variety of spatial scales.

Transforming place and space

Attention to the relationship between place and transgender is the topic of a
Gender, Place and Culture themed issue titled 'Towards trans geographies'.
Kath Browne, Catherine Nash and Sally Hines (2010, 573) argue that the
topic of gendered geographies has: 'focused on normatively gendered men
and women, neglecting the ways in which gender binaries can be contested
and troubled.' Within this issue Browne and Lim (2010, 616) describe trans
studies as

> a growing field of enquiry that seeks to redress both the absence of trans
> lives in queer theory (despite the conceptual deployment of trans sub-
> jectivities across this field) and the pathologisation of trans bodies and
> minds through surgical interventions and as mentally 'ill'.

Sally Hines (2010, 609) highlights the way the term 'trans' is used to include
a diversity of gender identifications. This diversity of gender greatly affects
how people experience different spaces at different times (see Doan 2007,
2009, 2010).

Geographers inspired by Judith Butler's (1990) book *Gender Trouble* are
disrupting the presumed naturalness of the binary man–masculinity and
woman–femininity. Gender – as understood through feminist, trans, queer
and poststructuralist theoretical lenses – is a regulated performance and
with this has raised the possibility of transformative power and politics
(Sharp 2009). A growing body of scholarship exists precisely to highlight

the shifting nature of gender, unsettle binaries, and examine the intersectional relationship between gender and sexual subjectivities (Johnson 2008; Longhurst and Johnston 2014).

Not all geographers are attentive to the mobility of gender identities (Browne et al. 2010). So while there have been several decades of critical feminist and queer geographical research intent on subverting dualistic thinking, it is impossible to fully break free from male / female binaries. The impact of gendered binaries cannot be overstated in human geography (Bondi 1992). Research in geography and beyond highlights the association of masculinity with the mind, rationality and so called legitimate knowledge. The other side of this binary – femininity – is associated with emotion, irrationality, and the body (Johnston 2005a). Many geographers, me included, have employed an embodied approach to geography scholarship in order to upset gendered binary thinking (Johnston 1996; Johnston 2005a). At the heart of this debate is the desire to unsettle and disrupt gendered binaries by taking embodiment seriously. Hence, geographers who consider the materiality and performativity of bodies expose the tyranny of binary gender (Doan 2010), the need for gender diversity, and a politics of inclusivity (Sheehan and Vadjunec 2016). This work, together, 'genderqueers' the discipline of geography alongside a critique of normative cisgender subjectivities, places and spaces (Johnston 2016).

When 'non-normative' sexed bodies inhabit 'normatively' gendered bathrooms, Browne (2004) notes, they may be subject to 'genderism'. Antagonistic reactions towards gender transgressions – or the 'bathroom problem' where women are mistaken as men (which I discuss in Chapter 4) – highlights the performative and material consequences of trying to live beyond binary gender norms (Browne 2006). In some spaces it is possible to be both, that is to be 'a right geezer-bird' (*geezer* is an English slang term for men and *bird* is slang for women (Browne 2006, 121). In other words, it is possible to be both and in-between man and woman. There are men with breasts that may feel uncomfortable in places such as beaches, swimming pools and changing rooms (Longhurst 2005). Certainly, trans men who are pregnant and lactating are another example of bodies in-between gender binaries (Longhurst 2008) as well as the excesses of body building and spaces of training (Johnston 1996).

One can find more transgender research when looking into the collective experiences of gender and sexually diverse lesbian, gay, bisexual, transgender, intersex, queer (LGBTIQ) communities. This grouping may have very different experiences of gender, making at times, an alliance between gender and sexual identity that may be awkward or even incomparable. Yet, many LGB people are 'gender outlaws' or at least challenge gender norms. The notion that 'we're stronger together' is helpful when engaging in advocacy work, seeking healthcare and so on. Sexual minority subjectivities, such as lesbian and gay, have captured the attention of geographers for many decades; it is only in the past few years that we see a growing visibility of bisexual and transgender research (Brown 2012). Acronyms such as 'LGBT', or 'LG-BTIQ' appear in many geographical accounts of sexuality, community and activist space. This cisgender and transgender grouping of subjectivities is designed to reflect collective interests 'and community as sexual and gendered minorities' (Nash and Gorman-Murray 2014, 757). The concern is, however, that merely adding the letters 'T' 'I' and 'Q' without paying attention to the difference that gender makes, is counterproductive. Therefore, it is useful to pull apart the acronym in order to understand relational identities and experiences of place and space (Nash 2010b). Careful consideration of the intersections of queer, trans and culture has sparked dialogue that compares place and identity (Aiello et al. 2013). Some geographers consider the way in which prejudice works to further marginalise transgender people within LGBTIQ communities (Valentine 2010). The grouping of LGBTIQ gendered subjectivities needs to be considered in planning practice and legislation in order to create inclusive queer spaces (Doan 2011, 2015). Such alliances can help create queer spaces that act as safe havens for gender variant people in cities (Doan 2007, 57). These 'queer' community spaces are important for the expression of non-normative gendered embodied, for transgender people as well as other 'genderqueer' people (Doan 2009).

'Gay villages' in cities are spaces that 'should' be inclusive (Hubbard 2012), yet even in Brighton, England, 'the gay capital of the UK' some gender variant and transgender people feel unwelcome, as shown by Browne and Lim (2010). Similarly, Browne and Bakshi (2013) found that transgender people are the most marginalised of the LGBTIQ grouping in Brighton. Tensions exist in this LGBTIQ activist communities (Browne and Bakshi 2013) and these may surface during gay pride festivals (see Chapter 5 in this book, and Johnston 2005b). It has become apparent that many queer spaces – entertainment spaces such as clubs and bars - serve as important haptic sites for constructing gender and sexual identities (Johnston 2012), yet they also exclude the 'queer unwanted' (Casey 2007; Tan 2013). So-called 'unattractive' people such as the 'drab dyke' (Browne 2007a) experience marginalisation in these drag performance spaces. Similarly, certain embodied expressions of gender diversity are both in and out of place at 'Pussy Palace', a Toronto women's bathhouse event (Nash and Bain 2007). Performance and politics come together on the stage where gender transgressions are applied to drag and queer femme movements (Dahl 2014). Drag king stage performances may sustain LGBT communities (van Doorn 2012) while some LGBTIQ spaces, such as those created by the Sydney Gay Games (Waitt 2008), also provide possibilities for inclusive transgender expressions and performances. While not completely eliminating transphobia, some stage performances may challenge the national imagination of contemporary Turkey to be accepting of gender variant people (Selen 2012). Queer and gay villages and bars (see Chapter 7), then, may simultaneously be dominated by gay men, yet also be a place of LGBTIQ gender diversity (Podmore 2013).

Collectively, and in many different ways, LGBTIQ community places and spaces may be both welcoming and discriminatory of gender variant bodies. A quick review of this research establishes the importance of political and collective alliances, yet too little attention is given to the gendered embodied differences of LGBTIQ community members, and transgender people in particular. It is politically important to hold onto acronyms such as LGBT or LGBTIQ, yet some people, because of gender expressions, are still marginalised or missing from LGBTIQ geographical inquires.

In the Queer West Village of Toronto, Canada, place is re-imagined thanks to a queer youth programme involving trans, intersex, questioning, and two-spirited people (Bain et al. 2014). Discrimination and marginalisation is evident when considering age. Not 'fitting in' is illustrated in the high levels of homelessness for transgender youth (Reck 2009), and this happens even in gay neighbourhoods such as Castro, San Francisco (Brown 2014).

Turning away from bars and clubs, and towards the altar, Andersson et al. (2011) consider LGBT lived experiences and theological views in New York City (Vanderbeck et al. 2011). Religion, genders and sexualities have the potential to transform Toledo, Ohio (Schroeder 2013) and create safe spaces for African American LGBTIQ people in Newark, New Jersey (Isoke 2013).

Gay, straight and transgender alliances on campus (Davidson 2014) or in women and gender studies classrooms (Kannen 2013) may help create gender variant educational geographies. At an even broader scale, and related to a nation's desire to create '100 world cities', the Indian Government – Shah (2015) argues – must consider its role and respond to LGBTQ activism and rights.

Transitioning at work, applying for work, and keeping work is being discussed by scholars (Bender-Baird 2011; Doan 2010). So, while there has been a shift from hiding one's trans identity to living it openly (due to the progress of transgender rights activists (Halberstam 2005)), discrimination and marginalisation still exist in many workplaces. Travesties – transgender sex workers in Brazil – prepare their bodies for work by using hormones and industrial silicone (Kulick 1998). The complexities of being travesti are also discussed by Williams (2013) who examines racialised sex work and exclusion in tourist districts of Salvador, Brazil. Silva has spent more than a decade supporting and researching travesties in Brazil, including those who move to Spain to work illegally. Her research emphasises the implications of illegal work, globalisation, and mobility (Silva and Ornat 2014). Payne (2016) notes that in Columbia right wing paramilitary forces, the state *and* leftist guerrilla groups are associated with transphobic and homophobic attacks.

Some geographers have illustrated the ways in which rural festivals challenge gender discrimination in rural places through the inclusion of drag queens (Duffy et. al. 2007). Also examining festivals, regional development and drag, it is worth considering lesbian comedic twins – Lynda and Jools Topp – who 'cross-dress to become 'Ken and Ken' (Johnston 2009). The

Kens' performances reinscribe a nation – Aotearoa New Zealand – as a place of the 'transgendered kiwi' (Brady 2012). Another semi-rural and politically conservative place – Trinidad, Colorado, in the U.S. – has become known as the sex change capital of the world due to the physician Dr Biber's successes with gender reassignment surgery (Johnston and Longhurst 2010). Browne (2009, 2011) ventures into rural spaces of the Michigan Womyn's Music Festival to consider the paradoxes and productive possibilities of womyn's space and Camp Trans. Many oppose the Michfest womyn-born-womyn policy arguing that it is 'essentialist', out-moded, and transphobic (Feinberg 1997; Halberstam 2005).

At the state level, research clearly documents pervasive discrimination against transgender people (for a U.K. example see Whittle et al. 2007). These country reports involve extensive consultation with transgender people and provide vital insights into the lives of trans people, and their treatment by government agencies and courts. They show, at the state level, that trans people are often not provided for in law and policy, and that inconsistencies and discrimination exist at the state level. The law often fails transgender people who are at risk of high incarceration and when in the hypermasculine spaces of U.S. prisons endure harsh conditions of confinement (Rosenberg and Oswin 2015).

Applying for, and obtaining, consistent documentation regarding gender identity can be a major hurdle for trans people globally, and raises a number of issues for migration (Seuffert, 2009). Nations assume their citizens – heterosexual and homosexual - are cis men or cis women who conform to 'natural' or 'normal' binary gender. Nations and regions may prepare responses for when natural disasters occur. Often overlooked, however, is the way in which transphobia impacts heavily on transgender and intersex populations in relation to natural disasters who have difficulties with, for example, evacuation shelters (Dominey-Howes et al. 2013), displacement and home loss (Gorman-Murray et al. 2014; D'Ooge, 2008).

The book pays explicit attention to transgender, gender variant and intersex people's embodied experiences in order to give insights into a myriad of structural inequalities, and resistances, produced and differentiated across spaces and places. Genders are relational and socio-culturally dynamic, and more research is needed to understand, specifically, gender variant geographies.

Some notes on methodology

Some chapters in this book rely heavily on research participants' narratives. For two decades I have conducted research on embodied geographies of gender and sexuality, starting with topics such as lesbian bodies and homes, women body builders and gyms, and bodily performances at gay pride parades. These projects appear in the book. More recently, I conducted a research project called 'gender variant geographies'. This latest research project, then, is just one source of information for the book, but it is an important one, so I outline some of the research process.

The primary method used in the gender variant geography research project was face-to-face interviewing and participant sensing. I conducted all interviews, and I draw on my participation experience of leading and being involved in activist groups and events in Aotearoa. In most cases, interviewees and I had a shared sense of place, experiences and identities. The interview data collected are from 22 participants – 20 from Aotearoa and two from the U.S. – and all were asked about how they identified and lived their gender(s) in various places and spaces (home, bathrooms, workspaces, healthcare places, leisure and sport places, travel experiences in places away from home, and nation states). Some people focused more on one place than another, and this was highly appropriate given the diversity of experiences. For example, for those who had not travelled, they did not have any travel experiences to discuss. They did, however, share ideas about where they might, or might not, travel.

Participants range in age from late teens to over 65, specifically I talked with: five people aged between 18 and 24; three people aged between 30 and 34, three people aged between 40 and 44, two people aged between 50 and 54, one person aged between 55 and 59, three people aged between 60 and 64, and three people who were aged 65 or older.

When discussing their stories, I use the gender identifier that participants gave me and I list them here to give a sense of the various ways in which people describe their gender: female transsexual; transgender woman; female; female MtF (male to female); female transsexual (MtF); woman MtF; transfemale; transmasculine; genderqueer; intersex; intersex trans male. Participants were also asked about their sexual identity, which proved to be equally varied, for example: bi, bisexual, lesbian, attracted to females, mostly gay but flexible, pansexual, queer, bi female, dates guys. I was not surprised that there was rich variety in the ways in which participants describe their gender and sexual identities as the participants, while small in number, represent a diverse population. Throughout the book, unless referring to a participant, I use the terms 'transgender', 'gender variant' or 'intersex' to be as inclusive as possible of all aspects of sex and gender with or without body modifications. Similarly, I use participants' gender pronouns (such as they, them, s/he, she, he). The use of 'they' and 'them' as singular pronouns shifts writing conventions and some readers may find this unusual to read. The use of singular 'they', however, is a vital part of gender inclusive politics.

Participants were asked to state their ethnicity and this produced the following range: European New Zealand; Kiwi; White; Pākehā;[2] White English; Māori; Singaporean; and Asian. I used my network of friends, colleagues, and 'LGBTIQ' groups to secure participants. Everyone was asked to identify their occupation: retired; accommodation manager; administrator, student, teacher, professor, baker, truck driver, unemployed engineer, security guard, retired train driver, support worker, textile designer, educator, counsellor, self-employed business manager. Two interviews were conducted via Skype because busy schedules meant that we couldn't find a time to meet in the same place. One participant emailed their responses to me. Most of the interviews were conducted at participants' homes. Sometimes they were conducted in a café or restaurant, an office, or my home.

Also important is the 'participant sensing' methodology I used when conducting this research, being an activist, and living in spaces that conform and disrupt gender norms. Engaging all senses – touch, smell, taste, aural, and sight – makes for rich encounters with people and places. This kind of sensory or haptic geography – 'bodies that touch places, places that touch bodies, and bodies that touch each other' (Johnston 2012, 1) – meant that I recorded encounters through bodily expressions.

Chapter outline

This chapter has charted some of the debates that exist about gender variant and transgender geographies. It represents the overview of what research exists, and provides a platform for this book. Each chapter will provide a deeper analysis of the spaces and places outlined above. Chapter 2 introduces readers to the embodied geographies of gender variance. The body is the 'geography closest in' (Rich 1986, 212) and in this chapter I consider the various ways in which one's gender variant identity is constructed, resisted, lived, and celebrated. Childhood spaces (and gender memories) are discussed, as well as the spaces and spaces of 'coming out' as trans and non binary.

Chapter 3 turns to the place that is considered most private – the home. For many years geographers have focused on homes, yet there is very little research on gender variance and home. Home is often the first place that one may express gender difference and start living one's identity. Home may also be, however, a place where such expression remains hidden from family and other loved ones.

In Chapter 4 I move on to – and trouble the distinction between – public and private spaces. This is the 'bathroom' chapter. 'Where to go' has captured the attention of media, politicians, schools, religious groups, and other institutions so much so that the bathroom has become an emblematic space of transgender, intersex, and genderqueer rights, reactions, and policing.

Chapter 5 focuses on spaces of activism and alliance. LGBTIQ activism and activists' embodied emotions can be powerful forces for positive social change and challenge cis gendered heteronormativities. They may also, however, reinforce hierarchies within and beyond activist spaces. The intersection of gender, sexuality and race is brought to the fore in these group events. Examples of transgender protests at gay pride parades are discussed and one particular protest, 'No Pride in Prison' in the Auckland Pride Parade, highlights the connections and fractures between LGBTIQ alliances and transgender rights.

In Chapter 6 I consider workplaces. Finding work, gender transitioning at work, and keeping work are discussed. The idea of security – both ontological security and work security – influences the ways in which transgender and gender variant people feel at work. I offer accounts of transgender people's feelings of (dis)comfort and belonging to illustrate the constitutive relationship between workplace, bodies and (in)security. In a time currently characterised as precarious and anxious, it is timely to examine the (in)secure geographies of bodies, lives and labours. Considering gender variant people's feelings reveals how power and privilege operate, and the possibilities of challenges to cisgendered workplaces.

Chapter 7 explores the difference that city night spaces make to gender variant people's lives. Specifically, I focus on entertainment spaces, such as night clubs, to think through the politics of drag performance. In these spaces we are reminded of the fragility of binary gender as well as the power of humour and entertainment to make people re-think cisgenderism.

Chapter 8 highlights the difference a nation – and moving between nations – makes to gender variant people's lives. Reports on citizen rights, legal status, and exclusions are used to give a picture of the vast differences that exist within and across national borders. The necessity for people's identification documents to align with their lived gender is discussed. Being able to move across the globe can create anxiety for many people whose gender identity does not match their official gender identification documents. I also consider the intersections of gender, ethnicity, social, cultural and legal status for fa'afāfine in Samoa.

The book concludes in Chapter 9 with an argument that exploring the multiplicity of genders points not only to the centrality of place in the production of gender variance, but also to the importance of gender and sex in creating specific places and spaces. In order to understand the production and expression of transgender, gender variant, and intersex identities it is useful to consider place and space. A geographical lens provides another way to deepen our understanding of vitally important issues.

Notes

1 The term Māori is commonly used to refer to Indigenous people – or tangata whenua (which literally translates to 'people of the land') of Aotearoa New Zealand. It was a term introduced by white settler societies to refer to all iwi (tribes) as one people. In reality, however, it collapses iwi differences when used to refer to all Māori in Aotearoa (Mead 2016).
2 The term Pākehā refers to people who are white and of European descent, and (usually) born in Aotearoa New Zealand. Pākehā is a standard term of classification in the New Zealand Census, yet it is a contested term. 'Kiwi' is a term used by Pākehā, Māori and other New Zealand nationals (Bell 2014).

2 'There's like a gazillion gender options now'

The title for this chapter is a quote from a research participant, Amelia, who describes the extended list of gender identity options offered by Facebook for individual profiles. Despite the many options, Amelia, a young transgender person, aged 18–24, NZ European, MtF, and pansexual, still finds coming out on Facebook a problem:

> I don't like Facebook. Facebook as a way of communicating between people is especially stupid for the LGBT community. You're looking at something that, you know, that you have to choose a name, you have to choose a gender ... you still have to choose a name, you still have to, either come out or make a fake profile. Both of which are kind of a nuisance. If I didn't have to come out I wouldn't, but sadly when you look like this you can't really avoid coming out.

This chapter explores the embodied geographies of gendered and sexed variant subjectivities. The diversity of genders span across, between or beyond categories of 'man' and 'woman'. It also spans across places. While much has been written on the multiple ways people express, present and live gender variant, or transgender identities (Hines and Sanger 2010), there are few accounts of the 'fleshy' geographies of gender variant bodies (but see Doan 2010; Nash 2010a, 2010b, 2011). Transgender bodies may be considered a space, as well as being in place and space. Adrienne Rich (1986, 212) calls this 'the geography closest in'. How we look, feel, and act depends on bodily shape, size, appearance, dress, health, age, ethnicity, sexuality and gender. To state the obvious, bodies are the primary means through which we connect with, and experience, other bodies and places.

Transgender scholars are calling for a focus on everyday lived experiences, or the 'unvarnished materiality of bodies' (Prosser 1998, 9). Henry Rubin (2003, 11), for example, argues that 'bodies are a crucial element in personal identity formation and perception' hence the fleshy materiality of bodies, including all sex characteristics, are central to gender self-identification. This focus on the lived, embodied and felt experiences of being gender variant is a response to gender scholarship dominated by theories. The complexities of 'an internal, persistent identity that is not in accordance with the biological body' (Cromwell 1999, 48) is illuminated through people's narratives. Embodied differences can be the basis of discrimination and prejudice, particularly when one's gender does not 'match' the sex assigned at birth. Butler (2001, 622), however, argues that the performative gesture of a legible gender is a 'presupposition of humanness' and 'governs the recognizability of the human'. In other words, the language of gender helps us understand embodied experiences. Feinberg (1992, 1993, 1997) brings together both discourses and lived experiences to focus on female to male (FtM) gender identities. Feinberg offers their own personal and embodied experiences in and through the performativity of transgender histories.

In what follows I first consider people's realisations of (not) belonging to normatively gendered spaces and places. Many participants invoke childhood memories, or significant places and spaces, where they gain a sense of their gender and sex. Feeling and finding one's identity and place was a common theme in my research. There are mixed reactions and I highlight the ways in which people feel both similar and different to others. Second, I turn to moments when individuals make space for themselves, 'come out' and start using a range of identity terms to define their bodies. Paying attention to bodies, spaces and language helps illustrate the multiple ways gender variant people feel in and or out of place. Throughout the chapter, gender variant identities intersect with other identities such as sexuality, 'race', ethnicity, class, age, and dis/ability.

Childhood spaces: gender memories

The recollection of childhood memories is a powerful way for participants to understand their current identities. In thinking about transgender identities nearly all participants told stories about feelings and experiences during childhood and adolescence. One participant, Sally, who is New Zealand European, female, and aged in her early 70s, recalls:

> Well I first started to notice something different when I was 12. The date is imprinted on my mind. My father died when I was nine, which is not unusual. Fathers often die young ... And I remember it really well and we shifted. We were in a house. We didn't have any money. There was no DPB [Domestic Purposes Benefit, a type of social security] or anything like that in those days, so it was just, you got on the best way you could and I think we managed quite well. And we had a rumpus room and in the room was a kauri chest [a trunk made out of a native timber] and I was thinking about it when I was coming here tonight. It was the right hand drawer at the top. And I remember there were a whole lot of women's clothes there and I dunno, I suppose they belonged to my mother.

> But I remember putting them on one day and feeling 'this, this is quite good' and not wanting to get out of them and I did that **many** times.... What am I talking about [was] 50, 60 years ago?
>
> (emphasis in original)

It is significant that Sally attaches the feeling of putting on women's clothes to the place she found them. With no family members around, Sally was able to secretly experience the sensation of women's clothes. Lucy, who is white, female (MtF), lesbian and aged in her 50s, recalls events from when she was four years old. Her memories, like Sally's, also include a chest / trunk:

> At age four, age four, um I had an older brother who is two years older and I had a girl cousin who was four months older than I was. We used to play together a lot. And we'd go to my grandparents' place in the country and there was an upstairs back bedroom that had this trunk full of wonderful old dresses and things and we would go up and play dress up. And my cousin's room, her mother and father had been divorced a few years earlier, and she was living with the grandparents and her mother and so her room was just down the hall so it was very easy to get there. It was just so much fun and we would put on all these dresses and then we'd go down to where the grownups were having drinks before dinner and we'd do a little cancan line [both Lucy and Lynda laugh at this memory]. And I remember one time doing that and it's just crystal clear, my grandfather recorded it on video, on whatever, on eight millimetre, I don't know what ... so anyway one time doing that I had a very clear sense that this was such fun and I just wanted to be a girl from the skin out.

Cindy, who is NZ European, transsexual, bisexual and aged in her 70s, recalls: 'Well my story goes back over 60 years but basically I was a cross dresser ... I knew from age seven that I had a thing for wearing female clothes and it just grew from there'.

As these accounts highlight, for the most part, recollecting childhood memories of gender difference is infused with happy feelings. They involve play – alone or with others – with clothing and performing gender. The risks of re-telling these childhood memories is that they may assert a type of gender / sex essentialism. Yet, there are risks of not re-telling these childhood memories as they speak to the embodied feelings – as Lucy puts it 'I just wanted to be a girl from the skin out' – that mark significant times and places. These memories show an understanding of the relationship between gender identity, place and bodily appearance from a very early age. For Lucy, she remembers this feeling at age four. For Cindy, it was at age seven. Sally recalls being 12 years old and finding women's clothes to wear secretly (discussed in more detail in the next chapter). As these memories are from childhood it is not surprising that participants remember specifics about home. Home is where one dresses and prepares for the day. Home, as discussed in more detail in the following chapter, is a place where one may feel less bounded by gender conventions associated with public space (Johnston and Valentine 1995).

It is here that participants learn that clothes are key and powerful aspects of gender identity. Dressing in women's clothes, when socialised as a boy and assigned male at birth, may be an act of gender resistance. Assumptions, therefore, about the relationship between biological sex and gender appearance are challenged 'from the skin out'.

When I talked with Grace, who is Māori, MtF transsexual, and aged 30, about significant periods in her life, she said:

GRACE: I have been this person since I was young. I officially start taking hormones at the age of 15. And how I got into that was I met two friends one day in Hamilton, and they told me about hormones. That's how I got into those. Yeah. It has been for a while now. I am 30 now. ... And it's just who I am really, yeah. I just felt normal, just the way I am.

LYNDA: And from what age do you remember that feeling?

GRACE: I think the feelings came on strong when I got closer to college [secondary school].

Rather than discussing childhood acts of resistance, Grace frames her memories as normal and every day. College, or secondary school, is the place where Grace's outward gender appearance became the turning point for her to seek and use hormone treatment. Going to, being at, and returning from secondary school increased Grace's awareness that she wanted to do something about her embodied identity. All participants negotiate and manage assumptions about the regulation of gender. Another Māori participant, Vonnie who is takatāpui, Māori, a transgender woman, aged in her late 50s, and works as an educator, said that when she was growing up she was not aware of any words used to describe transgender (in English or te reo Māori):

I knew I was a sissy cause I used to mince around, put my sister's skirt on or something. But I was just playing you know, not thinking you know, doing the highland fling. My mother used to say, 'take those clothes off now before I come and clip your ears! If your father could only see!'

Spade (2003) cautions against reliance on childhood memories of gender identity difference. In an article that critiques medical evaluations of gender identity for legal work, Spade analyses his own writing / diary that was done at a time when seeking a double mastectomy and construction of a male chest, or 'sex reassignment surgery', as doctors would label it.

'When did you first know you were different?' the counselor at the L.A. Free Clinic asked. 'Well,' I said, 'I knew I was poor and on welfare, and that was different from lots of kids at school, and I had a single mom, which was really uncommon there, and we weren't Christian, which is terribly noticeable in the South. Then later I knew I was a foster child, and in high school, I knew I was a feminist and that caused me all kinds of trouble, so I guess I always knew I was different.' His facial expression tells me this isn't what he wanted to hear, but why should I engage a narrative in which my gender performance has been my most important difference in my life? It hasn't, and I can't separate it from the class, race, and parentage variables through which it was mediated. Does this mean I'm not real enough for surgery?

(Spade 2003, 19–20)

Gender troubled childhood memories, therefore, have become medicalised and expected when seeking medical attention. Spade (2003, 20) goes on to say:

I've worked hard to not engage the gay childhood narrative – I never talk about tomboyish behavior as an antecedent to my lesbian identity, I don't tell stories about cross-dressing or crushes on girls, and I intentionally fuck with the assumption of it by telling people how I used to be straight and have sex with boys like any sweet trashy rural girl. I see these narratives as strategic, and I've always rejected the strategy that adopts some theory of innate sexuality and forecloses the possibility that anyone, gender troubled childhood or not, could transgress sexual and gender norms at any time. I don't want to participate in an idea that only some people have to struggle to learn gender norms in childhood. So now, faced with these questions, how do **I** decide whether to look back on my life through the tranny childhood lens, tell the stories about being a boy for Halloween, about not playing with dolls? What are the costs of participation in this selective recitation? What are the costs of not participating?

(emphasis in original)

Gender rules are learnt from a young age, and not just by those with gender troubled childhoods, as Spade (2003) notes. Strategic story telling of childhood memories is crucial for Kiran, who is aged in their early 20s, is Indian / Chinese, queer, and transmasculine. Kiran told me:

I'm the classic kind of non-binary story. The kind of 'born this way' story that people like to sell because as a kid I was, you know, three years old and fighting my parents because they tried to put me in dresses ... My grandparents were very, very unhappy because they wanted a beautiful Indian child they could dress up and now they were having a half Chinese child who didn't even want to wear dresses. So that um, it was pretty early on. I swung between, you know, refusing to wear powder-puff girls' pink stuff and refusing to touch anything except for, you know, this one pair of cargo pants that an aunty bought me cos she'd mistaken my gender because I had my hair short at the time ... Puberty kind of threw me into crisis I suppose because I hit it at like about seven and I had never felt like I was much of a girl but I think I took to puberty reasonably well, quite pragmatic with it, cos I felt like maybe it would give me more time to grow into a woman, right, because there was no other way to be and if I didn't fit 'girl' maybe I would fit 'woman'? But, you know, for a variety of reasons that never happened and I do have to wonder how much of that is related to the fact that basically there have been repeated incidents where, you know, they are going to teach you what being a woman means you know, and all that kind of girl stuff.

Kiran's reflexive account recognises the hyper-gendering of families and the heavy weight of disappointment shown by relatives. As Hines (2007, 53) notes,

> puberty may mark a subjective turning point in relation to an increased awareness of gender discomfort, the social and cultural pressure to live within the gender binary means that many participants worked hard at conforming to their ascribed gender role.

This was true for Kiran who, despite feeling they failed as a girl, worked hard to learn to be a woman.

Kiran's most formative years (their words) were in Singapore. They told me that they went to a school for 'gifted' children until they reached nine years of age. Kiran was unhappy at this school and when the family migrated to Aotearoa, Kiran faced the challenge that no one could understand their accent, at which point Kiran 'decided I was going to actually try my best to be a girl all the time'. I asked Kiran about this time when they were ten years old, and new to Aotearoa:

> I dunno, it was kind of pathetic … The natural group of friends that I gravitated to were girls who were very much like me, ostracised and not particularly feminine. You know, one of them was a girl who didn't speak that much English and we could communicate in my broken Mandarin at the time, and the other girl was quite fat and wasn't quite into sports and um she faced a lot of shit from everyone because kids are cruel. But I spent a lot of time following the popular girls around trying to learn what lip gloss was [laughs].

As Spade (2003) notes, it is not just gender that matters, but also class, race, and family. Kiran articulates and understands the various aspects of their childhood. Being of Indian / Chinese ethnicity, plus a migrant, meant they faced culturally prescribed assumptions about gender, yet also racism. Feelings of being 'ostracised' in Aotearoa led Kiran to create friendships with other people who were Othered due to being not white, not feminine enough, not slim enough or sporty. Friendships formed, then, through feelings of being 'out of place'. Jenny, who is female, MtF, aged in her early 20s, white, and mostly gay but flexible, said that she was bullied as a child at kindergarten.

She turns this experience around to reflect on how it 'strengthened her re-
solve' in the face of day-to-day marginalisaton:

> I first realised that I was trans in early December 2013. Before then I was
> heavily repressed, feeling wrong and out of place but without know-
> ing why. On some subconscious level I exhibited feminine traits that I
> couldn't really suppress and after years of trying to be someone I'm not
> I realised that I have no choice but to accept the truth that I'm trans.
> Upon doing so my depression became a non-issue and now I feel I can
> do things with my life. I haven't started hormones yet but I will be very
> soon … I suffer from multiple disabilities the most relevant being my
> cleft lip and palate. From kindergarten I was bullied consistently and
> I think the level of abuse I received in school over my childhood years
> helped me strengthen my resolve and get used to the idea that I'm dif-
> ferent and that people will look at me funny on a daily basis which will
> help the transition process immensely.

For others – such as Sophie and Mani who are intersex, and for Yann who
is intersex and trans – childhood memories about gender and place unfold a
number of secrets held by other people – parents, doctors, and entire communities. Sophie, who is Pākehā, aged in their early 40s, says:

SOPHIE: Right from early childhood I always had a desire to be more effeminate, or embrace feminine aspects and never had any clue why. Um I'd
never been told why I had operations. I knew I had an operation when I
was younger and I had to repeat it when I was 15.

LYNDA: How old do you think you were?

SOPHIE: The first one was four years old. I remember because I got a nice big
teddy and I've still got the teddy today. This is a reminder of the case.
My parents weren't told any more than the name of the operation and
the fact that it was for cosmetic reasons and to enable me to be able to
stand up and pee.

LYNDA: Really. So they didn't talk about it?

SOPHIE: This is back in the early 1970s and I think it's still a touchy subject with mum and dad, especially mum, especially as I try and talk
to her you can sense she's uncomfortable. Um and it's not that I blame
them because they didn't know, they weren't told, you know, so it's like
'there's nothing wrong there'.

Sophie talks about her sense of feeling different but not being able to understand why. She was assigned male at birth, then had operations to construct
her body as 'more male', yet very little advice was given to her parents about
how to manage and discuss Sophie's body. Sophie, as a child, yearned for:

> things like, these brochures would come out for school and talk about
> the girls' white socks, something as simple as that and I was like, 'I wish I

could wear them'. Um I would see the nurses and the nurse uniform, 'oh I want to be nurse' only for the fact that I want to wear a nurse's uniform. I just never understood. And some nights I'd pray to God [clasps hands and looks up] 'oh God when I wake up maybe you can change me' and of course even at that time I hadn't heard about transgender or anything. To add to the complication, I had a very conservative, Christian upbringing ... But as you grow up you just try and push these things out and just try 'oh just try, keep trying, act more boyish, get into sports and things'.

Pushing these feelings away, and working hard to conform to gender norms, meant Sophie's childhood spaces were often marked with confusion and no one to talk with. Occasionally, a media item might spark excitement for Sophie, such as a news item about a 'cross-dressing bar in Australia'. Sophie, however, pushed this feeling of gender difference 'back in its little box again'.

For Yann, who is Pākehā, aged in their 60s, intersex and trans, secrecy wasn't an issue as a child, but feelings of shame were:

I always knew about myself and knew about the surgeries and I was under medical care until I was about 13, I think. Um and I always knew that there was someone else so I wasn't alone but it was something that was very hidden and a shame.

Later in life Yann felt they were not alone because Yann's mother had heard about Mani. Mani Bruce Mitchell was born with a medical condition described as intersex and in the 1990s, Mani decided to 'explore and understand the secrets' of their childhood. Mani's (Mitchell 2015) aims are to

increase awareness about intersex and gender variance issues. I regularly lecture at a number of universities, speak, hold conversations in public. I have run workshops around the world and have been involved in the production of several TV documentaries, a film and a photography book.

Mani has also been involved in the production of a number of intersex television documentaries (Greenstone Pictures 2002; Ponsonby Productions Limited 2012). I spoke with Mani, who told me that their experience of growing up as a child 'was pretty horrific':

So we lived in this remote farming valley so trips to the town for us children were infrequent. I have very vivid memories because going to town was a big deal. So we all get dressed up in very fancy clothes. And then in the car taken to town and then everyone being car sick. And the dirt roads and the car full of dust and all that. But I can remember usually when we went to town, it always involved, for me, a trip to see the doctor. And I have my childhood memories. He [the doctor] seemed like an inappropriately prurient individual and there were always very intrusive painful physical examinations. There was no justification for that.

As a child, Mani learnt to 'disown' their body and 'never talk of things'. The intense medical scrutiny of bodies born not 'typically' male or female often creates feelings shame in adults (Feder 2011). Treatment that attempts to 'normalise' (correct or conceal) ambiguous bodies prompts feelings of 'isolation, stigma, and shame – the very feelings that such procedures are [understood to be] attempting to alleviate' (Preves 2003, 81). In another study, one young adult that Preves (2003, 65) interviewed says:

> The primary challenge [of being born intersexed] is childhood; parents and doctors thinking they should fix you. That can be devastating not just from a perspective of having involuntary surgery, but it's even more devastating to people's ability to develop a sense of self. I have heard from people that are really shattered selves ... The core of their being is shame in their very existence. And that's what's been done to them by people thinking that intersexuality is a shameful secret that needs to be fixed. So I think for most people the biggest challenge is not the genital mutilation, but the psychic mutilation.

Feeling and finding one's identities, then, are filtered through memories that evoke shame. Shame 'is a painful thing ... it gets into your body. It gets to you' (Probyn 2005, 130).

All participants employ strategies to manage childhood experiences and feelings that they either remember or discover later in life from other adults. Memories of childhood spaces and places, and subsequent gender management, surface throughout adulthood. These embodied and deeply felt identities are maintained, resisted, and performed in a variety of adult spaces, as the next section on places and spaces of 'coming out' illustrates.

Coming out spaces

'Coming out' is a term and practice often associated with people's non-heterosexual identities. For trans and gender diverse people, however, coming out involves different experiences. When some trans and intersex people live in accordance with the gender with which they identify, they may consider themselves successfully *out* when able to live life *without* disclosing their trans or intersex status (Goodrich 2012). This is also called 'passing' and refers to a person's ability to be correctly perceived as the gender they identify. Not everyone wants to 'pass', but some may, as Amelia notes at the start of this chapter: 'If I didn't have to come out I wouldn't, but sadly when you look like this you can't really avoid coming out'. This section explores participants' experiences of place and space in associated with their personal narratives about their gender identity. Participants also recount their own process of understand their gender identity, plus reactions by others to their gender. Strong feelings of attachment to place structure participants' coming out experiences.

Virtual spaces – such as social media and internet sites – are important spaces where gender variant, transgender, and intersex people search for information about bodies that transgress normative gender. Cindy said, upon buying her first PC, 'the whole world opened up to me when I typed in "cross dressing"'. She went onto say:

> It just blew me away. The number of people on there and the number of sites for cross dressers and from then on I joined up with a group called the 'Auckland Cross Dressers' I think it was and I put a note there: 'anyone from Hamilton wishing to have a get together?' And I got this reply from Anne, and Anne and I became great friends ... we used to go all over the place together.

Doan (2007) notes that trans populations have developed strategies for connecting across places, particularly when transgender populations are small in number. Over the past 20 years the internet has opened many doors (or 'the whole world', as Cindy puts it) for trans and gender variant people to connect with each other, particularly when coming out (Whittle 2002). Online spaces are where people can share information, exchange personal stories, support each other, organise groups and activism. Often, the first step for many people coming out is by making contact with a group via the internet (Doan 2007). Sally, aged in her 70s, however, felt overwhelmed by the internet when she was exploring her transgender subjectivity:

> If you go on the internet and you look up anything to do with transgender there's a whole lot of stuff there which is uncalled for, is sort of, is it, there's a whole business out there hankering to gay and lesbian, to alternative people. And I don't like the word queer. I find that queer isn't right. ... I don't like that word queer, I don't. For myself I find it, I don't find myself queer. Queer means something unusual, I don't find myself unusual at all. I find myself quite normal.

Rejecting the word queer, when making sense of oneself and coming out, may point to the desire to be gender conforming, as well as desiring a 'normal life' (Spade 2011). Cindy, also in her 70s, highlights some concerns about identity labels:

CINDY: Ah. I usually ssay female, transgender, and transsexual. I identify as transsexual. I can't get away from the fact that we have changed. I don't like a lot of these girls bitching and carrying on over their blimmin titles. It leaves me cold.

LYNDA: What are some of the debates?

CINDY: Well some of them don't like to be called trans. And they don't like being called transgender. I don't know. I'm happy to be calling myself transsexual. I don't care what every other person says. That's what we are.

My passport states I am female. My birth certificate states I am female. Now we can have our passport changed without having our birth certificate. So now we can have 'F' on our passport. That is a biggest change.

Previous research conducted by Katrina Roen (2001) found that tensions exist in Aotearoa transgender communities around 'passing' (that is, supporting the gender conforming binary) or not (criticising and subverting the binary). Some of these debates surfaced with participants in this study, usually with older participants.

Julie, who is a transgender woman, Kiwi, and aged in her early 40s told me that she didn't know how to describe herself:

Um I sort of still feel more just a cross dresser more than anything else because I haven't actually made any change. Still haven't got onto the hormone treatments with doctors and that kind of stuff, but hopefully that's not too far away from happening now. Um but I've always considered myself to be a female just in a man's body, ah but I can't sort of tell people I'm a lady, sort of thing. So I could try but it's like not a lot of people see that sort of thing. So you know it's a little bit vague. I'm not really sure myself and that's a little bit of my problem at the moment as I haven't got a great deal of motivation. Getting out of bed is quite hard in the morning and that kind of stuff cos I haven't really got anywhere in the last four months since I came out.

Coming out, then, also means coming to terms with labels, health, and negotiating a politics of identity. Julie lives in a small regional town and there are no specialist services to assist her with medical transitioning. Her mental health is suffering and, at the time of the interview, she was reaching out to other community members.

Another participant, Michelle, trans female, Pākehā, aged in the early 20s, is aware of how some bodies are constructed as gender conforming. She notes:

Right now I would define myself as, you know, quite a binary transgender. Um I can't say that in the future I will. I don't know. It seems that among a lot of people an incredibly common thing is that after they have transitioned they almost kind of drift back towards sort of the middle a little bit because I suppose … after you kind of come out to yourself suddenly you tend to sort of get into a whole bunch of extremely gendered things so to speak. Simply because it's kind of like making up for lost time. Um so kind of being feminine, or at least for me, a very, very feminine thing and all kinds of feminine things are all very exciting, even though in the future I'll probably kind of roll back from that a little bit as I kind of become more comfortable with it. Right now I would identify quite, quite clearly as a trans female. So quite binary, not really genderqueer or gender neutral.

Hines (2007) notes that there are significant differences in coming out experiences between older and younger transgender people. I too found this. Most of the younger participants were comfortable in online spaces, regular users of Facebook, Instagram, blogs and so on. Yet, as Amelia notes at the start of this chapter, she doesn't want to come out on Facebook and expressed her annoyance with Facebook and their long list of identity categories. She said 'You have to identify as something in order to have Facebook. You need to identify and that is where most trans people get scared because, you know, how do you identify?' Amelia finds this internet space, and the 'gazillion gender options' not helpful for managing her gender identity and prefers not to come out online. When I asked Amelia about her identity, she said:

> Well I guess I identify myself as a female. I'm transgendered, male to female. Um, I dunno. I fully identify as transgendered but I don't, you know go up to somebody and say 'hi, I'm Amelia and I'm trans'. That's just awkward. I just let them assume I'm female if I'm passing, and if they don't well … well you know that's what happens.

Amelia told me that she often gets addressed as 'mam', yet transitioning and coming out is difficult. She elaborates:

AMELIA: If I showed you me two years ago you wouldn't recognise the same person. I look completely different and I've come so far. But still just, I, I don't feel it when I look in the mirror … it was probably three to four months into hormones and before that, obviously, I did identify as male. I wasn't out to anyone. I was very secretive about it growing up. It was something that I kept guarded. It was a big surprise to my family and my parents when I actually did come out.

LYNDA: How do you feel about your body?

AMELIA: I mean when you grow up in a beauty culture it's really difficult having a body like mine and seeing everybody else around and feeling like, you know, there's is something wrong with my body even though I've come so far. It's kind of horrible but I think I'll get there. I mean I'm trying to lose weight and learn how to dress better and I'm starting to learn how to use makeup although I still look horrible in it (laughs). I can't get eye makeup right. It's difficult.

LYNDA: Oh you look great.

AMELIA: Yeah unfortunately, money is a huge issue. I mean, you don't really see it when you're going into transition. Well at least I didn't. I thought, you know, clothes can't really be that expensive because I've been buying clothes a long time and they weren't that expensive. But once I went to uni it was about having no money.

Here Amelia reflects on hegemonic feminine beauty standards, being poor, and feeling 'in-between' genders. When she is in public, she lets her body announce that she is a woman. This is also the case for others who find coming out and going out to be interchangeable. Hence, one's changing embodiment in public spaces forces a type of coming out without having to 'speak it' or use an identity term, as Sally explains:

It's strange, Diane [name of ex wife] became an anchor really. We go out together and I don't know, one day we will discuss this [Sally's transition]. It's just that we haven't had an opportunity. We go out to dinner together, we go shopping together, we go clothes shopping together. We do the things that two women do together. I'm much bigger than she is. She's about your size [referring to my small size] and I don't know how she reconciles that in her mind but she does and, and because of that I try and make myself as feminine as possible. I try and keep my mouth shut, actually. So I try and be as feminine as possible otherwise I'd sort of let her down a bit. Yeah that's a new development and I guess I'm lucky that I'm still going, moving, learning the role.

In these everyday spaces of restaurants and shops, Sally is appreciative of Diane who gives her confidence to be out and about in feminised spaces. Family life and intimate networks enable Sally to grow in self-confidence to the point that she can come out and stay out. Of a similar generation, Cindy, notes how important strong friendships are to coming out and staying out. It was the older generation that, prior to the internet, had to rely on newsprint and the postal system to find others, as Sally – who lives in a rural location – explains. At age 40 she subscribed and wrote to the *Penthouse Magazine*:

I always remember this magazine arriving and I remember reading it in the shed because I didn't show anybody. It was all secret, there's this secret life that goes on. And this is very common, you understand that? [I nod] And I remember opening up, and I couldn't believe the things I was reading. I remember standing there absolutely spellbound by this magazine. There are men dressed up in women's clothing, they were getting married. It was a sex magazine, that's what it was, but to me it was a revelation. I never, I just never thought about, so when was it? It was [when I was] 40, so 30 years ago. It would be about 1980ish, about 1980 sometime like that so homosexuality was still illegal in New Zealand. And magazines were totally prohibited and so forth and so on and now we have the internet. We go onto the internet and it's everywhere.

Sally also shared with me some of the spaces where she use to experiment with her identity: 'I guess that experimental stage lasted 'til I went overseas. I suppose I would seek out, which is common, prostitutes and pay to get dressed, not cross-dressing, much more than that'. In sex-positive spaces and away from Aotearoa New Zealand, Sally took the opportunity to try out aspects of her gendered identity anonymously. She felt unable to do this in her own country with a small population (in 1980 the New Zealand population was approximately three million people (Statistics New Zealand 2017)).

On a visit to the U.S., I had the opportunity to interview Emily who lives in Los Angeles. Emily identifies as female, MtF, lesbian, Asian, and is aged 40–44. Living in a large urban city means she had different coming out challenges. She describes her transition as 'easy' in that as a teenager she didn't go through puberty fully. She reflects:

EMILY: [This] is a good feature for someone like me but for a cisman is probably devastating, but for me I needed a different kind of treatment. So it's a condition that, that some people would not [sexually] develop, so in other terms, it's a failed puberty. Um, so swimming failed to 'cure' me, in fact unlike other sports where you would wear more clothes - which would make things easier – but swimming is very revealing and I lied to myself … I would try to hide in the water so nobody would see me, right? Cause the water would be up to here (points to her neck). I still have to get out of the water at some point.

LYNDA: And go to the changing room?

EMILY: Mmmmm yeah so … when I'm in the water I feel like a fish and away from this world.

The water is a space where Emily feels at ease and comfortable. It is the changing room and getting in and out of the water that is most challenging. Many participants discussed avoiding fitness spaces where one needs to reveal and expose skin, hair and bodily shape. Despite this, Emily said that she cherishes her early transitional (in-between) phase because it allows her to analyse structural gender inequalities:

Yeah I'm cherishing this phase while I'm still in it and I see things, but I am quickly losing my 'special' ability to see those things when they happen every day. So it's, I guess I mean to say, it's socialisation … I was parachuted in, like everything in, and the deeper I go into the transition process I suppose the more successful [I am], the less able I am to see these things. Like other girls and women, I begin to take certain things for granted, such as sexism. It's not good but somehow I couldn't have my guard up all of the time. It's really tiring. It's a very interesting process and kind of a battle like the pocket thing, right? [We both laugh at the thought of women's clothing not having usable pockets.] I thought of that and my friends were 'mm, right, of course but you know we don't think about it. It is the natural order of things'.

Emily analyses the everyday sexism she sees and feels, as well as the way dress normalises women's bodies. The current style of women's clothing is to not include pockets. When they are included, they are shallow and not designed to hold bulky items. If the beauty standard for women is slim and slender, then one's body is not supposed to bulge in places due to pockets. Emily's women friends had not thought about this until she brought up the topic.

> They've [her women friends] just accept it. Now I am used to the idea of grabbing my office key when I step out. There are no pockets in my clothes to put the key in. Sometimes I bring two purses, a small one just to stuff my keys and important stuff in like when I step out. A big one for books for when I teach my classes.

Other participants also discussed their 'in-between' and / or ambiguous gender. As noted at the start of this book, Yann is in an 'in-between place' and has been 'transing now for coming up ten years'. Yann doesn't identify as transgender yet embraces the flexibility and uncertainty of ambiguous gender.

As a child, Yann:

> was raised a female okay. Um, and I became a lesbian when I was 18 okay, and I'm 60 now. So I was a lesbian, I suppose I kept to that definition really until probably about 50. I think I was 50 when I started transing. After I came out as trans it was quite interesting. It took me awhile before I found a woman who would partner with me .... especially after I transed. I started to attract the femmes. Um, and ah, they ah wanted me in a role which actually I don't fit either. You know, I don't fit. I don't fit into a traditional butch femme category at all.

Jack Halberstam (1998, 142), in a chapter entitled 'Transgender Butch: Butch FTM Border Wars and the Masculine Continuum', asks 'If people who are assigned female at birth articulate clear desires to become men, what is the effect of their transitions on both male masculinity and on the category of butch?' Halberstam unravels complicated debates, or border wars, between FTMs and transgender butches. Masculinity, it is argued, 'is what we make of it; it has important relations to maleness, increasingly interesting relations to transsexual maleness, and a historical debt to lesbian butchness' (Halberstam 1998, 142).

There is a risk here, to theorise a type of free fall, fluid, and fragmented subjectivity without taking into account the realities, and negotiated power relations, that Yann lives with. Yann is no longer comfortable in lesbian spaces because their transitioning body is being read as butch lesbian. As noted by Halberstam (1998) Yann's experience has some resonance with some FTMs who first come out as lesbians before they come out as transgender. Yann provides further insights into the way in which their body is read:

> I think that if people view me now they first see me as a gay guy. People who don't know me well, um, there's the assumption that I'm a camp guy. So, um yes. And it's quite interesting now I'm partnered with a dyke. Someone who defines as a dyke. So, and that seems to fit better, actually.

Yann is aged in their 60s, is Pākehā, has a deep commitment to class politics, and over his life time, has experienced a number of radical shifts in the politics of gender and sexual identity in Aotearoa. This life experience provides a frame from which to reflect on the shifts in their gendered and sexualised identities. I now turn to some younger transmasculine participants – Joel, Tammy and Kiran.

Joel is aged in their early 20s, is Pākehā, and identifies as on the masculine side of genderqueer. Joel came out to their parents, two years ago. I asked how that went:

> Really good, they're really cool. Yeah they, they were quite surprised, very shocked. My mother, I think my mum was really worried about me in future relationships. That was her concern. And my dad didn't really, he was, yeah, they both have the view that as long as I'm happy they will support me with whatever decisions I make in my life. And it's really cool, like my mum now she kind of says that I'm more like me than I have been, like for the past like teenage years and all that sort of stuff. So it's quite cool. I'm very happy.

Joel shared the following about their identity transition from lesbian to trans masculine:

> I used to go along to a group, it was 'dykes on mics' kind of an open mic kind of thing and it was quite cool, there were lots of really cool people, lots of quite quirky personalities. I used to do their sound system for them. I came out to the person who organised it and we had a talk about whether it was appropriate for me to be there or not and so I stopped going and as part of that they told the rest of the group that I was transitioning so they would know why I wouldn't show up any more. Um and yeah I think there were a few people who didn't take it so well, out of that group. Um there was, I think they um, they saw me as some sort of traitor and weren't very nice about it.

There was little room for Joel in the space of 'dykes on mics', despite it being a place of 'really cool people' and 'lots of quirky personalities'. Like Halberstam (1998) I am attempting to understand the multiple versions of masculinity emerging out of both lesbian and transgender spaces. Joel became very involved in roller derby, to the point of becoming a referee. For Joel, the roller derby space feels more inclusive of non-binary gender identities and expressions. Joel also told me that they had been the trans representative for a queer youth group, and that the experience had been enormously rewarding.

In another interview – this time with Kiran – we discussed queer youth spaces. Kiran says:

> I tend to identify as trans masculine because a lot of that's about an inclination towards masculinity. And I've got some privilege in that sphere so that's something to acknowledge, [privilege] compared to

trans feminine people ... So that idea that masculinities are not nearly policed half as much as femininity is. I can't even write femmes transmasculine [on forms] cos at the moment, I'm not sure. I'm having this little crisis. I still don't know what ethnicity means, like, I actually still do not ... I came to gender identity through meeting the person who is still my partner now and they identified as transsexual and I looked that up and I was like 'there are more than two genders'!

In general, LGBTIQ places and spaces provide opportunities for the expression and performance of gender variance. Yet, and in contrast, some participants also discussed which LGBTIQ spaces and places they avoid due to feelings of displacement. Kiran went on to say that the youth group space was, however, free of transphobia because that are a lot of trans people there. Amelia also feels she can 'be herself' in this space. Yet, it is still a very white space, and 'the queer spaces at uni are the same'. The politics of transgender, quite obviously, reproduce other political struggles in these LGBTIQ spaces. While some transgender people many find strength in LGBTIQ youth groups, others like Kiran, find their identities and loyalties divided by their various affiliations. Halberstam (1998, 159) affirms that 'one axis of identification is a luxury most people cannot afford'.

In my next example I draw on an interview with Tammy. Tammy is 32 years old, identifies as transgender and genderqueer, and is Pākehā. For Tammy, context is everything, and they gave an example of going to a gas station in order to illustrate a complex 'in-between' space of trans masculinity. Ash says:

> I **hate** going to the gas station because petrol stations are so gendered. They feel really like everyone is going to pick either calling you 'Ma'am' or 'Sir' or even worse some version like 'Boss' or 'Chief' or 'Bud'. Like, I can handle 'Mate', cos Mate's fine, everyone uses 'Mate'. But like if I had to choose out of that kind of gender language, I rather would be called 'Honey' or 'Love' or 'Darling' or 'Sweetie'. But especially if you are at gas station and it's so male ... If they want to interact with me mostly they'll do some hard core **manning** like quite intense. They must be thinking 'You must be a man because you are probably not a girl because girls usually don't have facial hair' (emphasis in original).

'Hard core manning' here refers to the cultural context of hegemonic masculinity in Aotearoa where being tough is often celebrated as an ultimate manly attribute and connected to the sport rugby. Tammy acknowledges that in the gas station: 'some people who identify as transmen I think would find that really comforting and really enjoyable to be kind of indoctrinated in some kind of brotherhood. It is the opposite feeling to me'.

When gender variant people are constantly challenged about their gender identity, 'the chain of misrecognitions can actually produce a new recognition: in other words, to be constantly mistaken for a boy' as Halberstam (1998, 19) writes, 'can contribute to the production of a masculine identity'. It is important to address the ways in which trans people manage to affirm their gender despite the multiple sites in which that gender is threatened, challenged, denied, and violated. Tammy understands the ways in which gender is threatened, challenged and questioned when one occupies the 'middle ground':

> Well I am myself and constant, to some extent, within myself. But generally like one example is that now that I turned 30, I'm like, well, okay, it's been cool being on testosterone and generally having people recognise me more towards the boy end of the continuum, but that's starting to feel like just as much of a trap. Like it's starting to feel like the same problem that I had before which was that everyone was girling me all the time and there was no recognition of anything else, and no respite from any of that. And then now it's to the other end when there's no respite from people thinking I am a boy. Since I genuinely feel like I am in the middle, it's kind of ... It's difficult.

Tammy highlights the ways in which binary gender is mapped onto bodies and places. Being 'in-between', transgender and genderqueer means they are acutely aware of the social and biological construction of binary gender. The process of 'girling', as Butler (1993, xi) claims, occurs through relentless iterations that 'not only produce the domain of intelligible bodies, but produces as well a domain of unthinkable, abject, unliveable bodies'.

There are some similarities between genderqueer people, and those who are intersex and refuse to live at the edges of the binary. Mani, for example, spent 40 years trying to live within gender norms but this became impossible. Mani asked:

> Who am I? Who am I? Who am I? That's where this person emerged. So it's a person who is not fully male or not fully female but has elements of both of those. That's not constant. It changes. It's somewhat fluid. And that's when the reclaiming of my facial hair happened. So, for 40 years I have cut, in a quite shaming way, the facial hair. Cos females don't have facial hair ... I thought, what would happen if I grew it? Originally it was just going to be over a holiday, just as an experiment. It was so unbelievably powerful. It was like this one last place of my body where the difference was still there. I know it's not symmetrical and there's more hair on one side than the other, but I also kind of like that too ... And it became very powerful for me, here in New Zealand, I thought a lot about moko [Māori facial tattoo customs]. I thought about an image on your chin, that's on the face, that I am different. So it was around that but I also decided to change my name so Bruce was my original birth name. And that's interesting because when I did my research I went to

the births, deaths, and records. And then I think that I was probably the first person to do it. So they were very warm and open and I got literally taken into the archive to the 1950s records, and they showed me the original record, and my original name had been written in pencil and it had literally been rubbed out. It had been written in pencil and you could still see it on the page. So again that sort of active reclaiming, disrupting, I will take this name back. So Bruce my original name and Mitchell and then Mani, I don't know if I would do it now, but that's what I did back then. I was looking for a name that was both male and female. And in the English language, we don't have many names like that. Mani is an Indian Sanskrit name that means both male and female.

Names do things, and Mani is hyper aware of the power, process and practice of naming. I agree with Halberstam (1998, 173) who argues 'against monolithic models of gender variance' and I support the call for diversity.

I started this chapter discussing childhood spaces. In my last example, I turn to Tammy's reflections on how they feel about gender identity as they age:

who we are changes over time depending on our experiences, and what we are exposed to and what we think is possible and what we ... what the visions are. I talk about myself, my visions of myself, as an older person, and I really want to be an old lady. Like I feel like, the only way I can imagine myself as an older person, is as an old lady. Like that, that feels completely right and completely lovely. So you know at some point probably in the next however many years I will stop taking T because that would feel I think, like part of the process of moving in that direction again. Um, but it felt really important and really helpful to move in the direction of boy because that was kind of moving towards the middle at first. So, I think that my conclusion that I have come to now, is that I was always meant to be in the middle and it's just the matter of navigating the fact that the dominant culture I live in doesn't recognize that.

Acutely aware of anti-trans aggression, Tammy spoke to me about the steps they will take to ensure their personal safety against assaults and oppression. The lived experiences of 'being in the middle' of binary gender are also about age, affecting transgender and genderqueer bodies differently according to lifestages.

This chapter has canvased some of the lived spatial realities of what it means to embody, express and identify gender variance in different spaces. Most participants discussed childhood spaces when they first became aware of their embodied gender difference. The chapter also highlights spaces and places of coming out either through the adoption of particular identity categories and announcing one's gender, and / or sex, or by moving in and through places and letting one's body be read as gender variant. There are many divergent identity positions that can fall under the title 'gender variance'. Our bodies are also understood through other categories such as class, ethnicity, dis/ability and age. Rather than argue, wholeheartedly, for 'difference' and 'diversity', this chapter has sought to show how bodies and places are mutually constitutive *and* structured by power relations.

Some participants spoke of being 'in-between' binary gender while not finding many spaces and places where this is accepted, or indeed, encouraged. Some exceptions include coming out and being 'in place' at home (see Chapter 3 for further discussion), at support groups (see chapter 5 for further discussion), and in some virtual spaces.

By charting the lived, discursive and embodied experiences of childhood and coming out, this chapter provides snapshots of transgender embodied spaces. The complexities and (dis)congruities of gender variant bodies cannot be easily explained, rather, my intention is to focus on the 'practice and process rather than arrival at a single point of 'liberation'' (Spade 2011, 20). Our bodies – how we look, move, dress, and feel – matter to how we respond to other bodies, and vice versa. Gender variant embodied experiences cannot be separated from places and spaces. The next chapter – home – is a space that has been largely ignored by social scientists interested in transgender geographies. It is a space, however, that is highly significant to one's ontological security and feelings of (not) belonging.

3 Homes and familial places

Transitional spaces

> Four bedroom house with one bathroom
> Warm, sunny insulated old villa on Wilson St.
> Queer, Transgender, Vegetarian household.
> Large double room with good carpet, wardrobe, and wood burner/
> enclosed fire (queen bed, drawers, and large desk avail if wanted) …
> ***A home, not a party house***
> We are two feminist / politically switched on adults who work
> and have busy lives and come home to chill, one primary school kid
> who goes to school and spends every 2nd weekend away, and one cat
> who likes humans but not other cats or dogs. We want to live with
> someone who is relaxed, motivated, grown up, reliable, considerate,
> child friendly, LGBTQIA+, paying the board on time with no stress.
> Vegetarian or vegan. Smoking strictly outside. We don't want to live
> with a couple, a heterosexual person, or someone who is loud at night
> or drinks / does drugs / parties a lot. We also don't want to live with
> someone who is racist, sexist, homophobic, transphobic, fatphobic,
> hates sex workers, hates migrants, or is otherwise a jerk.
> (Trade Me listing for a flatmate/housemate, 21 January 2016)

The above excerpt from a 'room-for-rent' online advertisement caused a great deal of public interest in Aotearoa New Zealand. When the listing was shared on the social networking website Reddit (see www.reddit.com/r/newzealand/comments/424p52/no_heterosexual_people_allowed/), it quickly attracted over 300 comments and prompted other media items (see Suckling 2016; Thomas 2016). The debate raged between whether the advertisement was discriminatory or upfront and honest. Most of the comments bifurcate the debate into one of being straight or gay, and sidestep the importance of gender. Some cisgender heterosexuals were outraged that the advertisement marginalised them, claiming heterophobia and cisphobia (Thomas 2016). The country's Human Rights Commission criticised the advertisement (and bypassed the issue of gender), saying that, while it was legal to turn potential flatmates / housemates away based on people's sexuality, they encourage people not to discriminate. The result of this publicity drew a personal response, in the form of a 'reader report' from a member of Wellington's transgender community (see Name withheld 2016). The 'reader report' – by a 24-year-old trans woman living in Wellington – brings a transgender perspective to the debate, and highlights the need for homes to be safe spaces, particularly when home is one of the few places where one can be free from discrimination. The reader (Name withheld 2016) explains:

> I have no problem flatting with heterosexuals or cisgendered people but the necessity of safe spaces and queer friendly households is simply vital ... I transitioned almost two years ago, and had I not been able to find a group of liberally-minded friends to live with I would have stayed closeted until I felt safe. I've been asked to change my clothing by employers, physically and verbally harassed on the street and am still struggling to have my sex recognised on my birth certificate. In other words, having a safe home to return to can often feel like the only solace from discrimination at times.

Home is a key site for 'the construction and reconstruction of one's self' (Young 2005, 153). Social scientists and geographers in particular, understand the importance of the notion of home to subjectivity construction. Blunt and Dowling (2006, 24) recognise that 'home as a place and an imaginary constitutes identities – people's sense of themselves are related to and produced through lived and imaginative experiences of home'. Similarly, Duncan and Lambert (2004, 387) argue 'homes and residential landscapes are primary sites in which identities are produced and performed in practical, material and repetitively affirming ways'. In this research, home and subjectivities are posited as relational and ongoing. Young (2005, 140) adds weight to this argument, noting, homemaking is a process and 'give[s] material support to the identity of those whose home it is'. In short, home or subjectivities are not ontologically fixed, rather, they are understood as mutually defining and continually reproduced through the practices of everyday living.

Home, like all spaces, is both material and imaginative. It is a physical location and an emotional space of embodied feelings. Home is much more than a provision of shelter, it is material structure embedded with memories, emotions and meanings (Blunt and Dowling 2006). Often considered safe places, a home secures a sense of identity, agency and ontological security (Dupuis and Thorns 1998; Somerville 1992). Highlighting the disparities of ideals and lived realities of home, feminist and queer research shows the home is also a potential site of struggle, conflict, contrast, and as always a continuous process of negotiation (Blunt and Dowling 2006; Brickell 2012; Young 1997, 2005). The making of home at the scale of the house and neighbourhood 'operates as a site, source and process of resilience in heteronormative societies that are routinely discriminatory and potentially violent' (Gorman-Murray et al. 2014, 238). Geographers have noted that home is valorised as the site of heterosexual family relationships. Still today most houses are designed and built for nuclear heteronormative and cisgendered families (Blunt and Dowling 2006). Hence, the 'house as heteronormative haven' has been subject to sustained critique, particularly by queer and feminist scholars (Johnston and Valentine 1995; Gorman-Murray 2008; Oswin 2010). There is, however, a notable absence in geographic literature concerning the relationship between home and gender variant people's experiences. One exception is Choi's (2013) research, which considers the impact of gender policing in homes on transgendered people's embodied subjectivities and social interactions. Choi found that the home becomes an ambivalent closet space because of cisgender binaries and heteronormativity. This is certainly the experience of all gender variant people that I interviewed. Furthermore, most participants experienced socio-economic precarity, which meant homes were modest, shared, or – for periods of time – they lived without homes.

Responding to and building on the growth of ambiguous and contradictory analyses of home, this chapter uses a critical geographies of home approach (Blunt and Dowling 2006, 22), notably: 'home as simultaneously material and imaginative; the nexus between home, power and identity; and home as multi-scalar'. These three cross-cutting components structure the multiple realities of trans and gender diverse people's experiences of home (including not having one). Crucially, I argue that home (and homelessness) is a key means of expressing the lived realities of gender transgressive subjectivities in western contemporary contexts. The chapter draws on interview narratives, as well as contemporary media items, to show the complexity of gender variant people's relationship with home.

Gender variant homes: material and imaginative

Many participants evoke home not only as a physical location – where one lives – but also as an imaginative and metaphorical space of emotion and belonging. I asked Tammy, who is transgender, queer, Pākehā, and aged 30, 'can you talk about what home means to you, whether that is growing up in the family home or home that you currently in? Even the idea of home? What does it prompt for you?' They responded:

> Makes me think of the word 'nest'. I have been thinking a lot about this lately. Having moved back to Auckland and kind of, for the first time, the last couple of years of my life is the first time I actually haven't wanted to move, haven't wanted to travel and have felt really, really satisfied. So I have been thinking about this idea of like, what does it mean for me to be satisfied living in a place? Being home feels like being content. I am actually doing that. It's kind of, it still blows my mind to actually be in a place geographically and also in terms of this house and in terms of living with James and in terms of safety. Having other loved ones really nearby. All of that it feels like, wow, how did this actually happen so that I can be so peaceful? It's good. It feels good to be home.

Tammy's affirmation of home – as imagined and real – draws on idea(l)s of peacefulness and contentedness. As Tammy continued to talk, it became clear that the ongoing co-constructive relationship between gender subjectivity and home shifts and changes depending upon place and time. Young (2005) notes the importance of having control over one's domestic space. As such, for Tammy, acquiring and maintaining control over their home produces positive feelings and emotions. Various objects embody different facets of subjectivities – including inter-subjective relationships (such as sexuality, familial connections, spirituality, class, politics, and so on). Tammy elaborates on the importance of material objects:

> I think certainly that it's about furnishing the house, like you know, we are renting and having our own sense of aesthetics and what goes where and what feels good in what place. Things like having herbs, like pot plants. I really like having sage, rosemary and thyme and those kind of things around. My cat, my cat is hugely important to my sense of being at home. Dee, she is fantastic. She is definitely crucial. That's, those are really important things. Behind me [pointing to the back of the bedroom], I have a painting in my room. This is a painting one of my friends did. That's about fractals. And I have some quite deep meaningful feelings about fractals and that feels quite relevant to my sense of ongoing self and unfolding. There's that [pointing]. I like to have a little altar underneath that painting. I think I am quite a home oriented person in the sense of really wanting to make a cushy nest. Really wanting to feel like it's mine. I have been wanting, been committed to making things, like there is a lamp here that I re-covered in different fabric and put a moon on it. I like things that feel significant and important to me [and] that will be easily visible and available as I walk around my house.

I talked to Tammy via a Skype connection. They moved their laptop around the bedroom to show me significant items. Material possessions are important for Tammy because they embody their sense of subjectivity. The 'fractal' painting is symbolic of Tammy's ongoing and 'unfolding' sense of identity. 'Fractals' are never-ending patterns with a feedback loop. Perhaps the image resonates with Tammy's sense of gendered self because they spoke, at length, about not wanting to be fixed or 'trapped' into a notion of what it means to be transgender and genderqueer. Tammy is creating objects for their home – a lampshade, an altar – and these play important roles in connecting the fragments of gender, spirituality, and environmental politics. Items are juxtaposed and in the process, Tammy materialises a holistic identity at home (Gorman-Murray 2008). Tammy's material and imagined homemaking practices assert the legitimate presence of gender difference, reclaiming a rented Auckland home as a site that affirms and nourishes genderqueer subjectivities.

Gorman-Murray's (2006) research on the creation of separate home spaces for cohabitating gay and lesbian couples highlights the use of some rooms – usually bedrooms – for identity work. In opposite-sex households, he notes, too often the gendering of space links feminine and masculine subjectivities to particular 'workspaces', such as kitchens or garages (Gorman-Murray 2006). For Tammy, who co-habits with a partner, their separate bedrooms do important 'identity' work. The maintenance of their separate bedrooms allow for ongoing self-development and may also be a way to consolidate their relationship.

Another participant in rented accommodation – Cindy, who is trans-sexual, bisexual, in her late 70s, and New Zealand European – welcomed me into her home. Cindy lives on her own in a modest two-bedroom unit / apartment, which she has transformed with cherished burlesque-type possessions. One wall of her living room is a photomontage and contains numerous posters advertising burlesque shows. Many of the posters have Cindy's name included in the performance line up (see Figure 3.1 with us standing in front of the burlesque images). Cindy's living room is adorned with various costumes, wigs and accessories. The small space is transformed by the display of these items.

Figure 3.1 Cindy's wall of fame in her living room.

It became clear when I visited that Cindy that her home space is the link to her public performance space. Burlesque is central to her enjoyment in life and as soon as I arrived our conversation quickly turned from home to burlesque:

LYNDA: Just thinking about your home space and how important it is, now I can see just **how** important burlesque is for you. And it's fantastic to see you in all the pictures [on the wall]. I particularly love seeing you get lots of cuddles. And you are on the posters and have an amazing head dress.
CINDY: We all like to have our photo taken with the headliners. They have a workshop and we'll go and have a photo during the workshop. It's usually what we do. We go and get someone to take our photos at a workshop and I thought, oh I will print them out and put them on the wall. [Laughs] My wall of fame.
LYNDA: It's a brilliant wall, it's a brilliant wall.

(emphasis in original)

The wall, then, is a deliberate material statement of Cindy's gender – the images embody her love of hyper-femininity, of creativity, and performance. Rose (2004) suggests that what is done with photos – how they are arranged, displayed and viewed – is key to understanding the production of home space and in this instance, the production of home as a space of hyper-feminine transgender subjectivities. Cindy displays objects that reflect her gender performance, such as posters, photos, computers (to mix burlesque performance music) and speakers. Cindy's home space is – both material and imagined – a statement about her love of embodied gendered performance. Distinctions between domestic and public space collapse through the display of public performance images alongside the costumes hanging in the living room waiting to be worn.

A prominent intersex activist leader in Aotearoa, Mani Mitchell, introduced in the previous chapter, travels a great deal and as a consequence, their home (material and imagined) is vitally important:

My current home is very tiny. What's called a one bedroom [apartment]. The bedroom is a study and I have a bed in the main other room but it is very, very small and it's an absolute place of refuge and celebration. So as a child I was very drawn to [images of] gypsy caravans. They're very cluttered, colourful and over the top and that's probably my house … It's full of things that have real meaning to me. Things I have got on different trips, um, gifts that have come from different friends. Yet, the house is full of symbolic meaning. A very rich place. It feels like a very safe place. It's one place where I am absolutely authentic.

Mani's tiny home space reflects their sense of self, and as someone who needs a 'sanctuary' with the realisation that feelings attached to ideas of home are mobile. Mani continues:

MANI: What is interesting is that it's [home] somewhat mobile because, as you know, I travel. There are a few treasures that come with me. So I am able to symbolically recreate that [feeling of home] very quickly when I am on the road.

LYNDA: What sort of treasures do you take with you?

MANI: [Reaches for a bag] It is still in here. Because I have been so crazy busy since I was up in Auckland, I thought that last night, I haven't unpacked them. And I thought, there's no point in unpacking them because I am taking them … So this comes with [counselling] work and it's a crystal and it's very, very unusual. I don't know if you can see it? It's a titanium … And it has this little tiny bits of titanium held in it and symbolically, it's a heart. It was my learning to initially love myself. And what I love about this is the titanium. That's the strongest metal that we know of on planet earth, so it has symbolism for myself. [Holds up another item] This is an Indian hermaphrodite. A little figure. I got that in San Francisco and that has been with me. I got that at the first retreat. So you know the colour that I use symbolically is yellow for hermaphrodite. Can you see this...

LYNDA: I can see that little yellow shell.

MANI: And then this is just the rock that I picked up when I was doing the filming for Mani's story [see www.manimitchell.com]. Came out of Arizona. So that's what travels with me … And the importance for me is to be grounded and earth connected.

Indeed, Mani's relationship with home space is fundamental to their ability to move around the world and undertake intersex advocacy work. Mani's holistic approach to their embodied gendered identity is tightly bound with interpersonal connections and key life events. One reason why Mani's domestic objects are so central to their subjectivities is due to their personal journey of embodied discovery and a political desire to make a positive difference in other people's lives. Mani's home – its materialities and imagined notions of – travels around the globe.

Two transgender women that I talked with – Marlena and Steph – share a home as friends. Marlena, is a transgender woman, aged in her early 50s, and is of New Zealand European ethnicity, and Steph is aged in her early 60s, and is of white English ethnicity. Steph had to move away from her family, and this meant she was free to make her new home fit her gender:

MARLENA [ASKED STEPH]: What does change mean for you because you have to move away from the family and house? You needed to make your own space.

STEPH: Ah, it's a girls' haven, you know, makeup everywhere, there are clothes everywhere. I love it. I just … it's just like a dream come true. I love it. [Laughs]

MARLENA: I am a right messy bitch [Laughs].

STEPH: I could never find anything, but I can find my shoes, brush, powder. [Laughs]

MARLENA: It's cheaper living together. I do the laundry and she cooks in the house and we have a wonderful dinner. Wahoo.

Prior to this living arrangement, Steph didn't feel 'at home' when she lived with her wife and children.

STEPH: My family would go off camping [an outdoor activity involving overnight stays in tents]. And I would get there, then say, 'look I have had enough' and I would go home. The reason I went home because I wanted to spend time on my own and in a dress. And it's so powerful. I would travel from Timaru to Dunedin and go back again, just to get away from my family for a couple of days. Just to get two or three nights of being feminine, pretty and experimenting.

Steph's example highlights that one can live in a home, yet not feel at home. Her spatialised feelings of (not) belonging in her family home are in contrast to her current home that she shares with Marlena. They support each other and co-create a home where both feel 'in place' and not alienated. The materialities of home – and how gender variant people feel about home materialities – are closely related to the second of Blunt and Dowling's (2006) concept, the nexus between home, power and identity.

The nexus between home, power and gender variant identities

I started this chapter with a quote from an advertisement for a flatmate / housemate. The list of gender and sexual identities (which are favoured, which are not) clearly illustrates the connections between bodies, home, and power. Speaking about one's transgender subjectivity in relation to home is interconnected with family (usually parents) alongside other aspects of subjectivities, for example, religion. Many of the participants spoke of the embodied experimental possibilities that 'private' home space affords. In some instances, the nexus between power, home and identity is affirming to gender variant people. There are examples when this is not the case.

One of the younger participants – Joel, who is genderqueer, pansexual, Pākehā, and aged between 18 and 24, was living in their parental home at the time we talked. Joel appreciates the support they received when coming out as genderqueer:

JOEL: I'm living at home at the moment with mum and my three sisters. I don't see them very often cos I'm not home very often. Yeah ah, but it's quite cool. I yeah, they're, they're really supportive, really cool. My older sister has a boyfriend and he uses the right pronouns as well and is very careful about it and yeah, they're all very cool. And there's some things we talk about and there are things we don't. Like my mum's very squeamish so we can't talk about surgery or anything like that. Um, but she was, she had surgery a week before me I think, yeah a week before me, before I had top surgery so she came and visited me ... She hobbled into hospital and made the effort to come visit me which is, that was really sweet, yeah. And my dad, he usually lives at home as well but at the moment he's in Africa, he's been overseas for about six months or so. So that's, that's quite hard for our family, I think. But it's sort of very, yeah when we were all growing up that was quite common, he'd travel around (OK) so I was always the dad in the family. I always got to do the really masculine jobs around the house, so yeah,

LYNDA: So chop the firewood and?

JOEL: Yeah, yeah, [both laughing] yeah the things society says the dads have to do, so I really enjoyed it. It's a really safe space, which is very cool.

Joel and I share a joke about normative gender roles in homes. Many feminist and queer geographers draw links between gender, heterosexuality and domestic space, showing that gender is one of many crucial components of home (Robinson et. al. 2004). Joel's parents did not restrict their children's activities based on heteronormative cisgender roles, for which Joel was thankful.

Unlike Joel's feelings of gender freedom at home, Yann – who is intersex, transmasculine, Pākehā, and aged 50 – told me: 'I spent most of my life sitting on my bed reading books, except when forced to go out ... I think when I was at [my parental] home, my home was probably my bed and my bedroom. It wasn't necessarily in the house.' Yann went on to say that this helped establish a pattern:

> And so, therefore, when I went flatting [shared living], I had similar sort of arrangements. Um, but I bought houses quite often at the beginning. And I'd do them up, so I followed a family tradition there of doing up old houses. But I'd never stay anywhere long. I'm a, I don't stay at places long. Two and a half years is pretty much a long time for me to stay somewhere.

Mani, had ambivalent experiences of cisgendered norms as a child in their rural parental home. Within the privacy of home spaces, Mani was encouraged to do both 'boy' things, and 'girl' things.

> Growing up on the farm as I did, my parents were a paradox. Because of the privacy of the home they enabled this rather unusual child to exist inside the paradox. So I was someone that loved engines, knew the name of all my dad's tools in his shed. You know, I would also play dress up and it's in the Mani's story film. One particular Christmas I got high heel plastic dress up Cinderella shoes. I made them red and

they had glitter on them. I absolutely adored them for years. But I also got a one pound bag of nails. My brother and I both, because we weren't supposed to go into dad's tool shed. So we got given our own nails to build forts and trolleys with. I often thought about that you know. That was my parents really honouring me as a child and as I said they could do that in the privacy of the house, and were not able to do that outside the house. Now, I understanding that, I mean I understand socialisation and the period of time that was, the 1950s, and everyone was trying to be normal, whatever that was.

Here Mani doesn't fall into the trap of idealising rural childhood spaces. As Valentine (1997, 137) notes: 'Perhaps the most powerful imagining is of the rural as a peaceful, tranquil, close knit community … based on a nostalgia for a past way of life which is "remembered" as purer, simpler and closer to nature'. Rather, Mani is fully cognisant of the secrecy that surrounded their gender expansive rural home activities. Looking back to their childhood family spaces, Mani is able to make sense of their intersexed body.

In another part of the world, Emily, who grew up in a South East Asian country, describes her ethnicity as Asian, is a transgender woman, lesbian and aged between 40 and 44, elaborates on her parental home:

> Well um my parental home, like where I grew up, my dad is really Confucian, patriarchal and conservative. My mum is a dedicated Buddhist and I, other than my dysphoria and my private struggles, I have to say I'm relatively happy, but happy in a sense that I got no abuse at home but no real, how do I say, because when I came out to my parents last year they were saying to Carey [Emily's partner] and me 'oh my God but he was just so normal since we knew him and now he's like', and stuff. Carey got really annoyed and I had to say to my parents 'how would you know? You didn't even care'. Or 'it's different if your kid comes home and you say "how's school" and you look at the grades. You don't really have that deep kind of bonding'. Not to mention something like this. It's so difficult to talk about so I haven't talked about it [living openly as a woman]. Um I see my parental home as a castle of oppression. I see a place I wanted to escape. That's what was driving me, really, to leave my home country.

Emily's feelings about her parental home as a 'castle of oppression' and a place she 'wanted to escape' drove her decision to not only leave her parental home, but her home country. The multi-faceted dimensions of home – patriarchal, Confucianism, Buddhism, ethnic-cultural norms – was not a supportive space for her to live as a woman. When Emily opened up to her parents later in life, with her partner in support, Emily's parents reasserted ideas of 'normal' embodiment and the 'son' they had raised. Emily is thinking 'intersectionally' (Brown 2012; Hopkins 2017; Valentine 2007) and recognising the ways that individuals are actively involved in producing their own lives and homes. Her analysis of her parental home life overcomes some of the determinism of her parents' ways of thinking about gender identities that classify bodies into fixed categories.

A younger transgender woman, Amelia, aged 18–24, Pākehā, pansexual, talked to me about living at home with her family. Amelia has started to communicate with her mother, but not necessarily about issues to do with being transgender.

AMELIA: My mum makes a difference to home. Before I started to talk to her I used to trap myself in my room and play video games. I still, I still do that for the majority of my time, but I do actually, you know, come out and greet her every day and, you know, we talk about uni and her work. We actually talk although I still spend a lot of my time on my computer.

LYNDA: Do you talk about being trans?

AMELIA: No, no one talks about that. I would love to talk about it but I'm just too scared to raise the issue. It's a huge thing to talk about for me. I mean what do I say to them? I mean, I mean, my mum knows that I wanna buy more clothes and things like that but I'm too poor to discuss that. But I mean we don't discuss anything really. I don't understand why they don't want to but [pause] whatever. Families are weird.

Amelia's familial home experience, unlike Emily's, is structured around issues of money. Amelia stays at home because she cannot afford to live elsewhere (or leave the country, as Emily did). Amelia is aware of the household expenses and her own modest student allowance for university study means there is no money to buy clothes that would better represent her embodied gender. Importantly, Amelia is unable to voice the things that would make a difference to her life. In short, Amelia thinks her family home is 'weird'.

The thought of 'coming out' at home was discussed by most participants. When I asked Sally, a transgender woman, Pākehā, in her 70s, about her feelings and experiences of home she responded about wanting to come out, yet not being able to when surrounded by her family, six children, her wife and her mother.

Home was a place of not coming out. [Home was] a place to burn. Because you always allude to it, in hindsight, I would use things to test the waters. I mean, I never had the guts to come out. I had a lot of things to lose at this stage. I had six children, I had a wife, I had my mother. You know a lot of people were involved. Um so that continued on for a long, for a long time really just, it just continued on and on, yeah, it was just unpleasant.

Sally's lived experience of the family home – and being the 'dad' – powerfully demonstrates the policing of gender through cisgender heteronormativity (Choi 2013). Coming out and changing one's sex / gender proved to be too risky. As Sally notes, she had a lot to lose as her familial relationships centred on her being the 'master of the house' in a heterosexual family household. Sally, however, lived in a rural part of the South Island in Aotearoa and was surrounded by a community known for its difference, in other words, people who did not pursue or create wealth through capitalism, and where they could live sustainably by growing food, gain part time work and / or social welfare benefits. Despite this 'hippie-type' community of the early 1970s in rural Aotearoa, Sally stayed closeted about her gender. She told me about an experience, about ten years ago, when she met a woman who is a Christian, and this prompted Sally to try a spiritual pathway to help her resist living life as a transgender woman. It is within her home space where she 'tests' this spiritual connection.

SALLY: Quite recently, about ten years ago, I met somebody and she was a Christian and what I was doing [at the time] was I'd get a lot of [women's] clothes around – and I'd think this is abnormal, this is not the way to behave – and take them outside and put diesel everywhere and light them and they would all burn up. I must have burnt thousands of dollars' worth of clothing.

LYNDA: You would buy these clothes and then burn them?

SALLY: Oh yes, yes, yes did all sorts of stuff. Zips would be burnt, everything. But anyway at this particular time I became interested in Christianity or I became interested in my friend's God and I still am actually. And so on certain nights I'd go home and in a very loud voice, and I live in the country, I'd say 'Thank you God for a wonderful day. Thank you whoever you are. Thank you'. Because I find that, I find the universe or whatever it is up, or down there, or whatever it is around me, it envelops me, and I like to give thanks. So I'm giving thanks and of course, she was a Christian and so, and I really, I desperately wanted to become a Christian in different times of my life and I was being held up. I always remember, this is quite recent, going up to all my cupboards full of clothes, taking them outside and burning them all. Watching them burn and then going back inside and getting on my knees and then saying 'there you are God. I've done it. Let Christ come into my life'. Nothing happened. Nothing happened, hasn't happened yet, and it won't happen. But it's interesting that I have this feeling that, I had this feeling, that of guilt. As a Roman Catholic I would have been perfect.

It's difficult to convey the intensity of Sally's words in this written text. She spoke quickly, and emphatically, about her agony of wanting spiritual guidance. The burning of her dresses is an attempt to comply with cissexist understandings of embodiment (in other words, 'acceptance by God' will only happen if she denounces herself as a woman and accepts the male identity she was assigned at birth). As Sally questioned her gendered and spiritual self, her home space changed: wardrobes were emptied; clothes were doused with diesel; and, a bonfire blazed on her rural property. Sally's home, like herself, is an ongoing paradoxical process. Her home and her body are sites of conflicting subjectivities of simultaneously (not) feeling at home where tensions between her embodied identity and dominant hegemonic discourses of Christianity are lived out.

Grace also feels the tensions between home, power and gender identity. Grace is a transgender woman, aged 30–35, Māori, and lives in a small one-bedroom apartment. Grace talked about her parental home when I asked 'So what was home like for you? You mentioned your mum'.

GRACE: Memories of home. There are bad ones and good ones too. It was okay. I wouldn't say it was perfect. It wasn't perfect. It was like everybody else's life ... When I first transitioned it wasn't good. And then that [transitioning] made things tough so I left [home] at a very young age. And then I came back when I was, when I actually knew – properly – where I was with myself. And then I got, well, I kind of forced it onto my mum, to accept it [laughs]. And then she, after that, then she got her side of her family to accept as well. This is how I am going to be and yeah. I thought it was better for me to do it that way instead of just not having a family.

LYNDA: Yeah, it's very hard if you don't have family.

GRACE: Yeah because I know lot of girls like me that don't have support from their families. Even if they tried, even if they did try, their family will still push them away so my family was good in that area and they are still supportive of me now. Yeah, but my father, he's not [supportive]. Yeah, he finds it hard to accept. We don't have a relationship.

Grace's experiences of home are paradoxical and she has many transgender women friends whose families have pushed them away. The acceptance or rejection of a transgender loved one has the potential to significantly transform home experiences. I asked Grace if she had lived with her dad.

GRACE: Yeah for a certain period, that would have been early stages of, oh no, when I was about Form 2 [about 11 or 12 years old]. I was young and he would have left around that time. Just before I started college ... he doesn't accept me and that's why I don't want to have a relationship with him because of that and I don't think it's me with the problem. I think he has the problem. But he is Christian as well. He turned to Christianity to help change his ways, to help him, and he's very full on so I can't be like this. And like just certain things he says I don't get anything from him ... I have spent so much time without him now that I just don't really need him. My mum stands up for me, my other family stood for me and as long as I have someone, you know. Even if I didn't, I would still be me. But they have helped me a lot. I am quite close, me and my mum, my sisters and my brother, who is over in Oz [Australia] and I see aunties and uncles quite often.

Grace's home space is stretched out beyond the confines of her suburban home. Siblings, in Aotearoa and Australia, as well as aunties and uncles, provide a supportive 'home feeling' for Grace. The conflict over religious belief is a common relational identity theme that is played out in home spaces. Other studies (Whitley 2013) have found that some parents and siblings struggle to maintain their religious affiliations and identities while being supportive of their transgendered family member. The powerful influence of one's religious father, for Grace, is less important to her home space because he moved away from his family. She found support with other family members.

Julie, who is a transgender woman, Pākehā, aged 25–30, told me about her parental home. At times her parental home was a place where she had privacy, security and the freedom to try different clothing, but only when her parents where not at home.

JULIE: I've worn mum's wedding dress since I found it when I was probably 10ish, or something like that. I wore that dress hundreds of times and it got to the stage where I tore the zip trying to fit into it because I'd grown too big to fit into it.

LYNDA: Would you put it on when they were out and when you were home alone?

JULIE: Yes.

LYNDA: You think she realised?

JULIE: Oh she must have known. It got moved. I found it tucked away in the closet sort of thing and then it got moved a couple of times. It was right in the front and after a while it got put right to the back of the wardrobe. And then later on it moved and got put up onto a shelf and then it got put into boxes on the shelf. It got harder and harder to get to, sort of thing, and then it got to the stage where I couldn't fit it and tore the zip trying to fit it on. My mother was a bigger lady, but she was quite petite when she was married, put on a lot of weight after I was born. But I found a couple of other dresses that would have been hers from before she was married and that kind of stuff. So I wore mum's clothing from a very early age.

Julie's sense of her own identity, as a young woman, was formed in these home spaces where she would secretly try on her mother's wedding dress. She took a great risk, and in retrospect, imagines that her mother 'guessed' what was happening. Parental home spaces were also important to Lucy, who, like Julie, was sure that her parents 'knew':

LUCY: Mum would be throwing a night gown or something away and I would grab it and I'd experiment with lipstick. A funny story, one day I was trying on lipstick at night. I had a bedroom on the third floor of this house so it was far away from others, but far from the bathroom too. So

I tried to scrub the lipstick off and I hadn't got it all off and I went down to breakfast and my father looks at me, and … he was never very macho, but he sort of looks at me and goes 'what's that on your lips?' And so he spat on his paper napkin and rubbed my lips and said 'get all that off'. So now we know. And another time I had my mother's old nightgowns that I was sleeping in and I usually hung it up carefully and hid it, but that day I was late – as teenagers can be – to school and I think I just left it um tucked under my pillow and I hadn't made my bed. So when I get home from school that day my mum says to me 'I made your bed' [laughs]. Oh so you know, as soon as I could I went running up to [my room] and there's my nightgown hanging in the closest with my boy pyjamas hanging on top of it!

These periods of experimentation at home, stayed in homes. Lucy, Julie and Sally never talked to anyone about what it felt like to try on their mothers' clothes. It was in the micro spaces of bedrooms and bathrooms that they would dress and undress, try makeup, then hide all 'gender transgressive' evidence. Lucy had:

a box of things. There were clothes and little bits of makeup. Occasionally, um there was a *Playboy* [magazine] that had a description of a drag queen getting made up and so I got that and kept that … I was just sort of looking for any information about what it meant to cross this gendered line [and] how people did it. I was just fascinated but really, really repressed about it.

For Jenny, a transgender woman, New Zealand European, aged in her early 20s, home is a 'safe space where I don't have to deal with the outside world and can work on introspection and do the things I enjoy'. Michelle – also a transgender woman, New Zealand European, aged in her early 20s – was home schooled since the age of five because her parents were not happy with the education system, 'which is ironic because they're quite conservative', she reveals. Michelle was happy to be away from peer pressure at school, saying that being at home 'probably allowed me to question a lot more'. She explains that her parents, however:

Were quite conservative so in their minds it's quite clear, male – female, but at the same time as I was an only child growing up and because essentially the only toy I played with growing up was Lego – which of course is a very genderless toy – that means they didn't push any sort of gender rules on me, which probably allowed me to question my gender more easily, um which is also kind of ironic. But probably not desirable for them, so to speak, making life very complicated.

Michelle's parents placed great importance on home life. Michelle was home schooled by her parents, her parents worked from home, and home routines revolved around her parents' conservative Christian values and practices. Michelle found, however, that her home life was surprisingly free of gender rules. At the time we talked, Michelle had yet to come out to her parents but felt supported by peers and resolved to speak to them.

Another participant, Tammy, is positive about home as a site for friends and family:

> I have been also pondering lately about the intergenerational nature of home, so that's important to me. I feel like my home is the place where I live. I don't have kids and I don't have elderly relatives who live with us so it's really just like one generation of people, me and James at the moment. I would quite like to feel connected to an intergenerational space. When my brother has kids that would be nice. With my parents getting older, all of these kinds of things I would like to figure out ways to be more actively involved, make it feel more intergenerational. I still ... I feel like I have some of that in a queer context. Like our friends Calvin and Gene down the road have a one-year-old kid. James and I have been really, really close to Izzie as she's been growing up and we've been part of that support network for the new parents. That's been quite amazing so I think it's really important to me to have connection to lots of different parts of my family. And important that to some extent they are connected to each other like that. My parents have met a lot of my loved ones. And my loved ones know about my parents and know about my brother. That it's not compartmentalised too much. I think that was a 20s (age group) thing to do. It was very much like having these various friends and family but keep them separate.

Lucy had children to consider when she was 'coming up to transition' and this meant the home became an important space within which to explain bodily changes to her daughter (aged ten) and son (age eight):

> My daughter had found, before I transitioned, had found the wig. We had a bathroom with the back bedroom ... I had washed the wig and left it in that bathroom to dry overnight and the kids never went back there except this time she did and she came back and said 'what is in the bathroom?' This is my little ... ten-year-old, maybe, at the time and my son was there. And so I took them off to one side and we had this discussion and I said – this is my precursor to coming out – and I said 'oh that's actually mine' and they said 'but it's women's hair?' And I said 'well, you know, everyone has a little bit of male and female in them. You know your mum has a job and is very effective in her job and I'd like to you know I do most of the cooking and a lot of the child care and those are roles that are associated with women, and your mum does things that are associated with men and ... sometimes I like to wear women's clothes. No big deal but people won't understand if you talk about it'. So she got it and it sort of became our secret and I hate

burdening children with secrets. I talked to my therapist about it and they said 'that's okay, everybody has secrets it's not a big deal as long as you don't overdo it'. Um but that was in the back of my mind, but she was so accepting, which was fine.

Home, then, is a site of intergenerational negotiation, power and identity. As a taken-for-granted heteronormative and cisgendered space, it takes courage to come out as transgender, or intersex, as Sophie, who is Pākehā, aged in early 40s, and intersex, experienced:

SOPHIE: My sister is sort of understanding but is also cautious because her new husband has a son with someone else and they're worried about the custody thing. So she wants me to 'tone it down' when I go round [to her home], like no dresses. More for the fact that otherwise that could be brought up as the custody thing and they could lose custody of the son, so.
LYNDA: How do you feel about that?
SOPHIE: It's hard because you can't be the real you and you can understand sort of why they want to do it. But because of the kind of person that she is, she would easily jump on that chance and try and twist it to a kind of negative thing … Try and make it negative. So the best thing is just to listen to their wish and be careful.

Here Sophie conforms to her sister's, and partner's, wishes. When she visits their family home, she wears masculine, rather than feminine, clothes. Sophie understands the pressures and costs of gender conforming.

The domestic space of home and people's gender are mutually constituted. The home is a site of personalities and feelings of (not) belonging in which people's gender – as well as age, sexuality, ethnicity, and class – become meaningful. Homes are the spaces within which we say to others who and what we are. The nexus between home, identity and power is not bounded to the domestic sphere, but mediated by external or public social and political norms. Home is 'a contested territory: a meeting point between geopolitics and identity politics. [...] "House" and "home" are porous intersections of social relations and emotions, simultaneously public and private' (Brun and Lund 2008, 278–279). The agency and privacy of home – and gender variant people at home – is mitigated by dominant meanings and mores, surveillance, and state policies, as the next section shows.

Home and homelessness as multi-gendered and multi-scalar

The last section of this chapter is about the multi-scalar intersections of public and private, home and homelessness. Like all binaries associated with home (away / home, public / private, outside / inside, male / female, work / leisure, danger / safety) there are many disparities between the ideals and lived realities of home. Transgender, gender variant, and intersex people are aware that homes are not inherently private, but are exposed to external sanctions, therefore privacy and safety at home are constructed and continuously monitored (Gorman-Murray 2012). Home as both public and private (rather than either or) is well established (Blunt and Dowling 2006). Home is no longer idealised as a 'place liberated from fear and anxiety, a place supposedly untouched by social, political and natural processes' (Kaika 2004, 226). Home is always connected to the outside world.

In places that are considered to be 'safe havens' for LGBTIQ people, such as San Francisco's neighbourhood Castro, levels of homelessness are high for trans youth of colour (Reck 2009). Homelessness is one of the most visible aspects of domestic injustices and negative experiences of home (Brickell 2012). Homelessness, and 'unmaking' home, is 'the precarious process by which material and / or imaginary components of home are unintentionally or deliberately, temporarily or permanently, divested, damaged or even destroyed (Baxter and Brickell 2014, 134).

One of my participants – Vonnie, who is a transgender woman, Māori, in her late 50s, and works as an educator – told me about her home and homelessness experiences as she moved across the Tasman Sea, from Aoteaora New Zealand to Australia and back again. Like Indigenous peoples all over the world, Māori are over-represented among homeless populations due to ongoing processes of colonisation, racism and marginalisation. Vonnie was 17 when she arrived in Australia and soon after started living openly as a woman. She 'had no intention of coming home' to Aotearoa New Zealand and thought she would live in Australia for the rest of her life, until:

> I got really down and out and desperate and um I know what it's like to be homeless. Most have quite a simple view of us, but some of our sisters [other Māori transwomen], they had nice flats. They're only allowed to have so many people in their flat at one time. They couldn't open their flat to everybody. So, you'd sneak into their flat and sleep on the floor at night and then leave the next day so the landlords don't catch you.

Part of making Sydney 'home' meant also making money as a sex worker in Kings Cross, 'the **place** to be', Vonnie exclaimed (for further discussion see Chapter 6). It was impossible to 'go straight and get a job, and stay closeted' like she had been in Wellington, New Zealand. In Sydney, Vonnie had support from other Māori transgender women and it was here that she decided to gender transition. As a result of being supported and surrounded by her Sydney transgender whānau (family), Vonnie stayed with friends at Kings Cross. Home/lessness became a series of liminal spaces (Tunåker 2015) consisting of friends' couches and shared bedrooms in apartments. Work and life in Australia was both 'scary and exciting':

> I was with a whole lot of other older trans and that's when I started transitioning with them and just went to work on the street. Oh we were looking over our shoulders every night.

Also living in Sydney at this time was Carmen Rupe, a Māori transgender woman 'vivacious performer, businesswoman and brothel keeper, and LGBT rights and HIV/AIDS activist. She was a cultural icon in the transgender community who paved the way for many transgender men and women after her' (Engle 2013, no page number). Carmen, tired of harassment, left Sydney and returned home to Aotearoa in 1968. I asked Vonnie how she reached her decision to 'come home to New Zealand' and her response was sobering:

> It just came on. It just came on. And you just because you see your friends drop, eh. You see your friends drop and they die. You see them drop. Or they overdose and you're trying to revive them, to bring their heart back. Some get thrown out of cars and they get found on the street and all that starts to play up on your mind, eh. Then you just work for the money to just come home, to pay for your airfare and just come home, yeah.

Vonnie didn't want to 'come back in a coffin' as was happening to her sisters who, like herself, 'walked the streets all zombied-out thinking we were okay'. Coming home, however, meant she needed to inform her parents that she was living openly as a woman and in the late 1970s, this meant Vonnie had to hand write and post a letter:

VONNIE: I wrote to them. I explained to them that I'd changed, blah, blah, and sent them some photos. Oh yeah, sent them some photos. And I said to myself, 'Oh well, here goes, they know now'. After writing then I waited and waited and nothing came. Finally, mum wrote to me and said 'you were born to me as a son, you'll be always my son no matter what'.
LYNDA: Really?
VONNIE: Yeah, so, okay. And I said, 'Oh that's fine' and they were from the old school so that was quite hard. My father was an army man, you know from World War II and everything was so staunch and my brothers were really staunch too so they all went to the army.

Vonnie was aged in her late 20s when she returned to Aotearoa. Her father 'wouldn't have a bar of it', which is a colloquial way of saying he wouldn't acknowledge her as a transgender woman. He didn't, however, turn her away. Vonnie was able to stay in her parental home, which is the house that she continues to live in today. A key aspect of Vonnie's acceptance back home in Aotearoa, was her work in the marae [Māori meeting house], which 'wasn't very welcoming but I just kept on going, and they got use to me', she said. On one occasion, during a wānanga [meeting] to learn waiata [song], Vonnie recalls:

> I entered and sat there listening to all of them talking and then that's when I stood up and said 'What about a girl like me? What is their duty or what do they do on the marae?' And it went all quiet, no one looked

at me and um my auntie sort of brought it up the other day 'cause she still remembers me standing up saying that and everyone was sort of shocked at the time. They knew it [transgenderism] was around ... This prominent old man stood up and said, 'while you wear a dress, you are a woman' ... so that gave me more encouragement. So he knew, and he was just about hitting 100 at the time, but he must have known how things were before [colonisation]. Yeah, but it's just um something that he wasn't even shocked about so he knew what was coming. I got an answer from him and now they let me stand up and do karanga [a call made by women to summon visitors onto the marae]. I call all the visitors on if there's no one else there. I just don't do all of it. When I've got older women, aunties there, I let them do it but they're getting a bit lazy so they say, 'Go on you go do it' [laughs] 'You do it, come on'.

Being accepted at home, on Vonnie's marae, means being part of marae ti-kanga [protocol] and taking part in welcoming visitors to the marae. This is part of a Māori understanding of home, which is linked to Indigenous identity and whakapapa [ancestral links]. Elizabeth Kerekere (2017, 81) explains:

As Māori we claim our identity through whakapapa over countless generations of ancestors. Whakapapa places us within a whānau [family], hapū [sub-tribe] and iwi [tribe] which in turn connects us to marae and 82 specific tribal areas on Papatūānuku, our earth mother. Because of this, whakapapa is central to takatāpui [gender and sexually diverse Māori] identity and spiritual connection to tūpuna takatāpui [gender and sexual diverse ancestors]. It is clear that fluid sexual intimacy and gender expression existed among Māori in pre-colonial and post-contact times and has continued ever since. It was accepted without punishment and in spite of repressive English measures.

The space of the marae is crucial to Vonnie's feelings of being 'at home'. Following and leading marae protocol is a powerful statement of acceptance for Vonnie.

Another example of home as multi-scalar is expressed by Tammy, who is 32 years old, identifies as transgender and genderqueer, and is Pākehā. Tammy refers to an imagined neighbourhood as home. Ideally, Tammy wants to live closer to town, because: 'I've some um very gender cool friends there as well. Cos it's very cool having [cisgender] people who are supportive, it's slightly different having people who understand. So that would be my ideal.' A trans-friendly neighbourhood (as opposed to a gayborhood (Brown 2014; Doan 2007)) is important for Tammy because it would mean living in an area where other trans people understand the embodied joys and challenges of trans' experiences. Tammy's imagined home – as an important site of trans identity and power – is framed as a multi-scalar, porous space (Blunt and Dowling 2006). Constructed through embodied emotion and the need for support, the ideal home includes the ideal neighbourhood. Queer-friendly community spaces, then, maybe a home away from home. This is the case for Michelle, who told me that one particular organisation:

> The Rainbow Youth Organisation has kind of been my home away from home essentially also because it is just up the road from university as well … I'm kind of there as much as possible. Um it's a really terrific, terrific place.

Home can exist in many forms and in many places. This chapter has focused primarily on the home as a private, family or extended family / whānau place. Transgender, intersex and gender variant people have inequitable access to home and senses of 'home' operate differently across diverse populations. Home is an important site to understand gender variance. The design of homes, their symbolism and social functions, can tell us a great deal about how they affirm and / or discourage gender transgressions. As Doan (2007, 62) argues, although transgender people may find some feelings of home, and level of safety in gay or queer spaces, they may also face difficulty as 'in most overtly gay spaces there is little to no visible gender queerness or any indication that such variance is tolerated'. Nevertheless, gender nonconforming people sometimes do create or locate spaces for themselves (Nash 2011). As I explain in the next chapters, a sense of home may be developed further through community places, in commercial venues – bars and clubs – or within rural spaces, and within and across nations.

4 Public and private (in)conveniences

For a transgender teenager, something as simple as going to the loo at school can be a huge stress. So two Wellington schools are leading a dunny revolution: fitting gender-neutral bathrooms for students who feel uncomfortable using 'male' or 'female' bathrooms. Wellington High School has transformed its level 4 boys' bathroom into, well, just a bathroom. And Onslow College is soon to follow suit, spending tens of thousands converting an old block of girls' toilets into gender-neutral facilities.

(Edwards 2016, no page number)

These two schools, mentioned in a news media website, are joining a growing trend of schools moving towards 'de-gendering' bathrooms, that is, creating bathrooms not specifically assigned for either boys or girls, or men or women. Rose MacKenzie – a Wellington High School student – says: 'Some people don't identify with male or female fully, so it's hard for those people not feeling they can go into one of those bathrooms' (Edwards 2016, no page number). Rose, who is 15 years old, said that she sometimes avoids using bathrooms in public, not knowing which to choose: 'If I go into one I know I'll be told this is the female bathroom, but if I go into the other I might receive threats because of, you know, what I look like' (Edwards 2016, no page number). Rose is part of an LGBTQI+ student group. The group – led by Ed Smith – provided the school board with a comprehensive proposal to re-fit existing bathrooms, which involved converting urinals to cubicles, fitting sanitary bins in each cubicle, and changing signs to read 'bathroom': 'The board of trustees listened to what the need was and within a short amount of time it was all done and dusted. It's kind of a boring story, in a good way' (Edwards 2016, no page number).

The establishment of gender-segregated public bathrooms has prompted transgender and gender nonconforming people to have to face even more harassment, violence and discrimination. Private bathrooms are also spaces that are under surveillance and where gender is disciplined, normalised, and / or transgressed. It is a mistake, therefore, to think that public bathrooms are gendered and that private – in the home – bathrooms are not.

The seemingly mundane spaces of home bathrooms are places of gendered performances, bodily maintenance, surveillance and self-policing.

The micro-spaces of toilets – in homes, bars, cafes, train or bus stations, public restrooms – may be sites of refuge and/or sites of gender confrontation. Each site takes on spatial, social and cultural norms. The persistent problem with toilets is that they are usually based on binary gender segregation and this cultural phenomenon is literally built into places and spaces. All humans need to excrete, yet toilets do not include all genders. In contemporary western societies, public toilets are one of the very few explicitly segregated spaces. Biological differences between men and women are used as justification for toilet design, which in turn naturalises the gender binary into a social and cultural norm. The ways in which public and private toilets are provided and maintained reflects the dominant discourses of time and place, hence, social inequalities based on gender, 'race', class and sexuality are displayed on the walls and spaces of 'ladies and gents' toilets.

The first part of the chapter highlights existing scholarship on gender and bathrooms and addresses the question: why are bathrooms gendered? The gendering of toilets has always been a contentious issue and I include historical accounts of architectural transitions of public and private toilets, alongside changing understandings of gendered, sexed and classed bodies. Theoretical frameworks, such as the heterosexual matrix and gender performativity (Butler 1990, 2004a; Namaste 2000; Prosser 2006), surveillance (Foucault 1978), and abjection (Grosz 1994; Kristeva 1982) are offered in order to make sense of the relationship between sex, gender, sexuality and bathrooms. In the second part of the chapter, I draw on interview data, as well as media sources and stories, to examine normative and transgressive gender experiences of bathroom spaces.

Why are bathrooms gendered and segregated?

Gender-segregated public bathrooms were first established in Paris in the 1700s. As part of the requirements for hosting a ball, a Parisian restaurant constructed '*cabinets* with *Garderrobes pour les hommes* [literally 'cloak room for men'], with chambermaids in the former and valets in the latter' (Wright 1960, 103). These new spaces not only regulated bodies – along the lines of gender, and sexuality – but also were designed to demarcate the upper classes from all others. The Victorian era's obsession with class, heterosexuality, and genteel respectability alongside social anxieties about women in public places meant that British and European public facilities exaggerated gender differences (Cavanagh 2010). Men's bathrooms were built into the fabric of public city spaces while women's public toilets only started to appear in the late 1800s and early 1900s (and often tucked away at the back of 'Department Stores'). At this time the introduction of women's 'lavatories' was met with great objection as the prevailing discourse was that women should not be in public places and hence should only use toilets within their homes (Gershenson and Penner 2009). Class identities played a significant role in these attitudes and 'ladies who shopped promiscuously mixed with factory or flower girls – presuming, of course, that the latter could pay the facility's prohibitive (and also controversial) penny charge' (Gershenson and Penner 2009, 6).

Toilets have also been sites of racial segregation. For example, white workers in a power company 'Western Electric' in Baltimore, Maryland, demanded segregated toilets between white and black workers (Gershenson and Penner 2009). These bathroom anxieties highlight the supposed threat of mixing with the Other and the white fears of catching sexual diseases from black people in integrated bathrooms fuelled many protests (Boris 1998).

What is now understood as the common toilet, Cavanagh (2010, 28) notes, arose from:

> the management of excretion in London, along with the technologies of the water closet developed by a host of engineers, plumbers, and inventors of the eighteenth century, led to a historically unparalleled privatisation and gendering of the elimination function. The present-day problem of restroom segregation by gender is attributable in no small measure to the curious ways Londoners and Parisians imagined and gave rise to toiletry provisions in city spaces.

This so-called private, bounded and gendered body was produced by London's sanitation reform and modernisation development, which associated binary gender order to conventional values such as health and longevity. The city's engineers and planners, concerned with the threat of cholera and the way it spread by sewage contaminated drinking water, constructed spaces in the city that were purified by inscriptions 'Ladies' and 'Gentlemen' and gendered bodies were channelled into separate toilets and compartments. Bodies were quarantined according to their biology. This gender purity, 'its intelligibility and segregation by type of genitals, was associated with health and well-being – longevity, sanitation, and protection from disease' (Cavanagh 2010, 40).

On the other side of the globe, closer to my home and in the 'frontier' city of Dunedin in Aotearoa New Zealand, Cooper et al. (2000) examine the making of gendered citizens and public toilets. From between 1860 and 1940 'public toilets, and their expanded and feminised form, rest rooms, became sites within which bodily excreta and exchange – urination, defecation, menstruation, lactation and sex – could be licitly or illicitly carried out "in private in public"' (Cooper et al. 2000, 417). The gendering of bodies was similar to that in London and Paris, that is, due to the rise of sanitary engineering and architecture in the West from the eighteenth century. The obsession with public health was driven by medical hygiene, but also part of cultural and social formations about 'public decency' (Stallybrass and White 1986). Urban spaces 'modernised' with new technologies, what it meant to be 'respectable citizens' and which became entwined with departures from grime, human waste, and the industrialised city. Discourses of disgust and respectability, therefore, were gendered, raced, classed and written onto bodies.

It is worth pausing to consider further the gendering of disgust and respectability. Julia Kristeva does this through the notion of abjection in her book *Powers of Horror* (1982). She considers the horrors that various bodily boundaries and orifices evoke. Kristeva critically examines the conditions that culturally produce so-called 'proper' bodies. By 'proper' she means obedient, decent, clean and law-abiding. She also argues that 'abjection' is the cost of producing the 'proper' body. Abjection:

> is an extremely strong feeling which is at once somatic and symbolic, and which is above all a revolt of the person against an external menace from which one has the impression that it is not only an external menace but that it may menace us from the inside. So it is a desire for separation, for becoming autonomous and also the feeling of impossibility of doing so.
> (Kristeva 1982, 135)

That which is abject is something so abhorrent that it both attracts and repels. It is both fascinating and disgusting. The abject 'does not respect borders, positions', exists on the border and is 'ambiguous', 'inbetween', 'composite' (Kristeva 1982, 4). The abject is what threatens any notion of fixed or stable identity. It is neither good nor evil, inside nor outside, but something that threatens the distinctions themselves. The promise of binary gender is established by abjection. Other gender and queer theorists have built upon Kristeva's abjection, particularly Butler (1990, 1993), Grosz (1994), McClintock (1995), Thomas (2008) and Young (1990), to understand how certain bodies and places become Other, devalued, and marginalised in late-modern societies.

Much of this scholarship draws on the work of anthropologist Mary Douglas (1980). Douglas examines the connection between dirt and corporeality. Bodies are symbolic of social order and that which is associated with 'rational' systems, hence bodily fluids such as urine, faeces, semen, milk, blood, tears and sweat represent potential threats to social norms as they upset so called acceptable social order. Following from this, anxieties about pollution and purity are actually ontological anxieties about order and disorder, borders and crossings, being and not being. Bodies become 'dangerous' when orifices open and leak onto other bodies and surfaces. There is nothing inherently 'dirty', rather it is matter out of place and hence the power of dirt to threaten social orders.

Some bodily fluids cause horror and disgust, while others may cause little anxiety. Kristeva (1982) extends Douglas's ground-breaking text *Purity and Danger* (1980) to develop sexed and gendered notions of abjection. The abject prompts fear and fascination because it threatens the distinction of binaries such as Self and Other, male and female, heterosexuality and homosexuality. Douglas does not pursue the relationship between abjection and sexual difference, rather it has been Kristeva (1982), and other French corporeal feminists (Irigaray 1984) who argue that sexed and gendered difference is at the core of power relations and distinction between clean and dirty. The clinging viscosities of some body fluids are usually associated with femininity, while solidity and firmness are associated with masculinity. Grosz (1994, 207) highlights the threat of abject bodily waste to ontological security:

> Excrement poses a threat to the center – to life, to the proper, the clean – not from within but from its outermost margin. While there is no escape from excrementality, from mortality, from the corpse, these do not or need not impinge on the everyday operations of the subject or body. The (social and psychical) goal is to establish as great a separation as possible from the excremental, to get rid of it quickly, to clean up after the mess.

In a book called *Queering Bathrooms* Sheila Cavanagh (2010) argues that the obsession of gender segregation of public toilets is a response to perceived threats to sexual difference and to heteronormativity. Gender and sexuality are closely related in heteronormative ideologies, or the heterosexual matrix, as Butler (1990) calls it. To identify a person as gay or lesbian, one must know in advance if this person is male or female. Therefore, in an attempt to maintain fixed sexual subjectivities (hetero, homo, lesbian, bisexual), genders must also be fixed. When, however, these subjectivities are queered, Cavanagh argues, bodies become linked to dirt, disease and disorder:

> The contemporary toilet is a place where gender variance and homosexuality are linked to dirt, disease, and public danger. Those who are recognizably trans are subject to persecution for using the 'wrong bathroom' in ways that are not only callous and cruel but compulsive and curious. The urgency with which one seeks to clarify the gender identity of another or to expunge gender-variant folk from the public lavatory entirely is beset by worries about disease and disorder that, in the present day, are overlaid by angst about a racialized and class-specific gender purity.
>
> (Cavanagh 2010, 7)

Bathrooms, and other spaces where bodily boundaries are broken, are considered 'dirty' spaces and only a few geographers have considered them (Browne 2004; Longhurst 2001). I have used the notion of abject when understanding the gendered politics of women body builders and gyms (Johnston 1995, 1996, 1998); gay pride parades (Johnston 2001, 2005a); and food (Longhurst et al. 2008). Geographers have mapped out some of the relationships between place, space, bodies, bodily orifices, and bodily fluids (Johnston 2006, 2009; Longhurst 2001, 2005). Clara Greed (2003) examines public bathrooms from an urban planning perspective and notes that the toilet is not designed for women. Greed (2003) advocates for inclusive public toilet designs that will allow ease of access when one has a pushchair, for example.

Kath Browne (2004) uses the term 'genderism' to describe the hostile reactions and discriminations that occur to gender ambiguous bodies in bathrooms in the south of England. The research focuses on '"women" who are read as men and those who do not identify with either sexed category both of whom confront the necessity of defining oneself in relation to dichotomously sexed sites such as toilets (in this case women's toilets)' (Browne 2004, 332). Browne (2004) extends the work of Jack Halberstam (1998) and Sally Munt (1998, 2001) on female masculinities, to highlight the policing of gender transgressions in toilets and the strategies her participants engage in to subvert the binary spaces of bathrooms. In the remainder of the chapter I too highlight the politics of policing gendered and sexed bodies in bathroom spaces.

Legalising exclusion: policing bodily norms and spaces

I started this chapter with a quote about bathrooms in two Wellington secondary schools in Aotearoa New Zealand. There are also bathroom challenges in other secondary schools in Aotearoa, as Stefani Muollo-Gray found out. Stefani, the first out transgender student to attend Marlborough Girls' College was 'called to several meetings with teachers because she used the girls' toilet' (Eder 2016, no page number). The school cited reasons of 'safety and comfort' as their justification. Stefani was aged 16 at the time and started a petition addressed to the Aotearoa New Zealand Education Minister, asking for access to the girls' toilet. The political lobbying and pressure paid off, and Marlborough Girls' College Principal, Jo Chamberlain, reached the decision that 'any student at the school, including Ms Muollo-Gray, could now go wherever they feel comfortable, whether that be self contained gender diverse toilets or larger bathrooms for either gender' (Keogh 2016, no page number). This is a positive step towards respecting all expressions of gender in Aotearoa's school spaces. It would be a mistake to say that this acceptance is sweeping the country; rather, these architectural and social changes are happening in isolation. At present, 'there is no cohesiveness across schools or across the country and no way that that's been enforced or reviewed by [the] ERO [Education Review Office]' said Duncan Matthews, the executive director of Rainbow Youth (Enoka 2016, no page number). Oliver Rabbett – who started transitioning from female to male when he was 16 years old – prefers gender-neutral bathrooms:

> Having gender-neutral bathrooms would be a massive step forward, he says. 'I would definitely have used them in school, though I hadn't fully transitioned'. Oliver is now 21 and uses men's bathrooms, yet he prefers to use unisex bathrooms, if available.
>
> (Enoka 2016, no page number)

Schools, universities and other large institutions are often understood, and legislated as, public space, even when they may be privately owned. Doan (2010, 643) names the public restroom in her workplace – a university – as one of the 'scariest spaces for a person in the midst of a gender transition'. The most 'private of public gendered spaces', for Doan (2010, 643), 'risked discovery and a potential confrontation with others outraged by my perceived transgression'. She goes onto say that in the U.S.: 'for trans people, the full weight of the legal system is against us, requiring a hyper-vigilant approach' (Doan 2010, 643). When Doan advised her employer about her gender transition, she was told to use the disability restroom, a single-access room on a different floor to her office. Doan (2010, 644) reflects:

> I was informed in no uncertain terms that they had received legal advice that I was NOT to use any restroom designated for women until such time I had undergone complete gender reassignment surgery and provided documentation of a court ordered change of sex.

Even with the law on one's side, bathrooms are risky spaces for Doan (2010, 644):

> Since my surgery I make a point of using the women's restroom wherever I am, though I do my best to do so unobtrusively. Because of my large body type, I am still liable to undergo the kind of genderism ... but since the law is now on my side, I am willing to risk it.

Across the globe, various countries have legislated to regulate transgender and gender variant people's access to public bathrooms. These 'Bathroom Bills' influence people's access to restrooms based on a person's sex assigned at birth, or the sex that is listed on a birth certificate. A prominent Bathroom Bill, the Public Facilities Privacy and Security Act in North Carolina (and officially called An Act to Provide for Single-sex Multiple Occupancy Bathroom and Changing Facilities in Schools and Public Agencies and to Create Statewide Consistency in Regulation of Employment and Public Accommodations, yet more commonly known as House Bill 2 or HB2) was passed in March 2016. North Carolina's legislation prevents people from using government-run public restrooms and bathrooms that correspond to the gender they identify with (Gordon et al. 2016). This legislation means that transgender people who have not taken legal or medical steps to change the gender assigned to them on their birth certificates have no right to use bathrooms of the gender with which they identify. Furthermore, people who are genderqueer and whose gender category is not enshrined in state legislation, face discrimination and marginalisation if challenged when using a public restroom.

Not surprisingly, there has been a great deal of protest against these Bathroom Bills. The European Union was quick to condemn them, issuing a statement on 12 May 2016:

> The recently adopted laws including in the states of Mississippi, North Carolina and Tennessee, which discriminate against lesbian, gay, bisexual, transgender and intersex persons in the United States contravene the International Covenant on Civil and Political Rights, to which the US is a State party, and which states that the law shall prohibit any discrimination and guarantee to all persons equal and effective protection. As a consequence, cultural, traditional or religious values cannot be invoked to justify any form of discrimination, including discrimination against LGBTI persons. These laws should be reconsidered as soon as possible. The European Union reaffirms its commitment to the equality and dignity of all human beings irrespective of their sexual orientation and gender identity. We will continue to work to end all forms of discrimination and to counter attempts to embed or enhance discrimination wherever it occurs around the world.
>
> (European Union 2016)

Dean Spade – an associate professor at Seattle University School of Law, founder of the legal collective Sylvia Rivera Law Project – experienced first-hand the severity of gender-segregated bathrooms when he was arrested for using a men's rooms at Grand Central Station, New York City. He reflects: 'I spent 23 hours in jail on a false trespassing charge, catching a glimpse of what more vulnerable trans people (homeless, youth, people of color, disabled) face daily' (Spade 2003, 17). The experience is captured in an article called '2 legit to quit' published in the zine called 'Piss&Vinegar' (www.makezine.org/2legit.html). Another consequence of the arrest – and perhaps more traumatic – centred on 'trans legitimacy' and 'image maintenance' with members of trans communities blaming Spade for the arrest because he failed to meet a particular kind of 'manliness', or even 'transmanliness'.

Some of my research participants were eager to perform gender in line with normative understandings of masculinity or femininity as a way of resisting their parents' expectations. This was the case for Emily, who sought legal documentation of her gender transition to try to ensure safety when using public bathrooms in the U.S., but also to go against her mother's wishes. Emily, who is in her early 40s, a transgender woman, and Asian, recounts her mother's reaction when she told her about starting her MtF medical transition: 'My mum … when I told her [about my gender transition] she said um "oh I'm happy for your gender but don't use the women's restroom". … So the first thing I did was to complete the legal changes'.

Emily talked to me about her bathroom anxiety: 'It's very unsettling and after you pass the point of no return, where do you go?' She continues: 'I'm definitely not comfortable in the bathroom, and I know a lot of people wait for surgery and stuff, especially facial surgery, not so much bottom [surgery], to use the women's bathroom'. Physically changing one's body to align with one's gender identity can be done through surgical and hormonal therapies. Gender reassignment surgeries (GRS), also known as sexual reassignment surgeries (SRS), is a term used to describe a set of procedures designed to change the physical appearance and function of bodies. 'Bottom' surgery usually means the surgical alteration of a penis or vagina in order to resemble that of another sex. Facial surgeries, also known as facial feminisation surgery (FFS) alter the 'typically male facial features to provide a more feminine appearance' (Ainsworth and Spiegel 2010, 1020).

Despite securing her legal documentation that shows she is a woman, Emily still feels uncertain. She asks: 'Where am I supposed to go? 'Cause all right there are gender-neutral bathrooms but they're not easy to come by … I already feel really unsafe as somebody might force me to use the men's bathroom'.

Earlier in this chapter I made the link between 'abjection' and bathroom spaces. Not only are bodily boundaries broken and resealed in bathrooms (prompting fears based on the disruption of clean / dirty, order / disorder binaries), they are also places where binary gender is troubled. The anxiety and fear experienced by gender variant and transgender people in relation to bathrooms is linked to social and cultural norms about what and who is deemed to be in and / or out of place.

Joel, who is aged in their early 20s, Pākehā, genderqueer, and pansexual, spoke to me about their discomfort at secondary school when it came to physical education (PE) classes. Joel says:

> I used to do PE and it used to be I'd walk in the [women's] changing rooms and go directly to the bathroom to change, and then I'd walk out. You had to look directly down the entire time and it was quite hard but you get like kind of comments um, like other people feeling uncomfortable cos you're there, cos they think you're looking at them. And it's like no, no I'm really just trying to leave.

For Joel, the closed single cubicle bathroom is the safest space away from others in the adjoining changing room. Joel's embodied gender is seen to depart from social and cultural norms at school and other students react in a way that links gender identity to sexuality. In other words, students assume Joel is sexually attracted to women because of Joel's genderqueer / butch embodiment. In this school space, gender recognition is mediated through heterosexual matrices (Butler 1990). During this interview Joel told me about other people's experiences:

> A really good friend of mine, he had his top surgery a couple of months before me and he had a lot of trouble um, they forced him to use the women's toilets and wouldn't let him into the men's area.

Joel's transgender male friend was in hospital for surgery to construct a male chest (another type of GRS), yet was unable to choose the bathroom that best aligned with his identity.

One of Cavanagh's research participants – Neil, who is genderqueer and transmasculine, experiences reactions that collapse gender and sexuality into a heterosexual matrix:

> *That you need to go pee has nothing to do with what you do in a bedroom, but for some reason, it has everything to do with what you do in the bedroom.* Because you enter a female or male space, it's assumed that … you like the person in the [next] room and when you contest [gender] category[ies], then automatically your sexual orientation comes into questions … Straight away it raises a whole issue with homophobia; it's weird how they are interrelated in a bathroom.
>
> (Cavanagh 2010, 174, emphasis added)

If the bathroom has become a hyper-heterosexual space, then it is expected that one's gender conforms to heterosexual norms. Modern public bathrooms subject all bodies to become objects of other people's gazes so that our gender is under scrutiny every time we use a bathroom (Foucault 1978). When a body is deemed 'out of place' in the bathroom, gender variant people may adopt a variety of strategies. Halberstam (2012), a well-known queer theorist and someone who lives in-between binary gender, notes:

> I still use women's restrooms and I avoid any and all contact on going in or coming out. If someone looks frightened when they see me, I say 'excuse me' and allow my 'fluty' voice to gender me. If someone looks angry, I turn away but mostly I just ignore what is going on around me in the restroom and do what I am there to do.
>
> (www.jackhalberstam.com/on-pronouns/)

Julie – who identifies as a transgender woman, Kiwi, is in her early 40s, and at the time we spoke, was seeking work as a boat builder in Aotearoa – avoids gender-segregated toilets:

> I don't, at this stage, I don't actually use the men's or ladies' toilets. Public toilets and stuff are avoided. At the mariner, which is one of few places that's got men's and ladies' toilets and it's got a public toilet which is either, so at this stage I'm just using that.

Julie told me: 'I would get up [out of bed], as a female, and use the gender-neutral marina bathroom, walking out there in public as a female. Then I would change and go to work as Calvin [a man] because I wanted to keep my work at that stage. She has a long-term goal to use the 'ladies':

> When my boat goes back in the water there's a possibility that I might end up using the showering facilities. I'm not quite sure. I'll have to ask if I'm allowed to use the ladies. I'd rather use the ladies than the men's. But I don't know what's going to happen with that.

Julie questions whether she will be allowed to use the women's bathroom (with showers) due to others reading her embodiment as beyond normative gender binaries. The marina is a semi-private space and subject to rules created by the marina corporation. Julie's employment and housing was precarious. She was finding it difficult to gain, and keep, a job and hence, have enough income to pay rent and marina fees. She was repairing her yacht and had a loan that she could not repay. She was also estranged from her parents as familial relationships had broken down when Julie told them she was transgender. Julie, and other homeless people, do not have private spaces so are constantly reliant on public bathrooms. All of these stress factors meant Julie's health was suffering.

For transgender people with employment, negotiations with employers about bathroom use can be long and frustrating. In the U.S. state that Lucy lives in, all high school bathrooms are deemed to be public and therefore part of the state law's about gender-segregation. She explains the implications of this law:

> State law, in my state, required that they be sex segregated and I could not use the bathroom until such time that I had legally had gender reassignment surgery and had gotten a court order legally changing my gender. So that's possible to do but you know the surgery itself is quite expensive and I am blessed that I have some resources and was able to pay for this but many trans people can't. I have a dear friend who's had an orchidectomy, which is just castration, and ten years ago she couldn't use that surgery as adequate for changing her gender papers. But in the interim – between when I had my surgery and she had hers – they now allow that surgery (which is not really gender-confirming surgery, so you know, um she couldn't go to a public bathroom and pass) but she can use that surgery to get her paperwork changed. That's huge, that's really changed. But when I did it, that was not yet acceptable.

There are many intersecting discourses here that complicate people's use of gender-segregated bathrooms. Changing laws, changing bodies, and 'passing' as the gender one knows oneself to be, are carefully negotiated. For Lucy, and for many others, being able to use a bathroom, at her high school workplace, meant meeting legal requirements. A number of participants told me that their employer requested they use the 'handicapped' bathroom, which is also designated 'unisex'.

The reproduction of the binary gender order continues in my own workplace, a university, when new buildings are designed that continue to build gender-segregated toilets. The 'bathroom problem' is an ongoing agenda item for the university's Rainbow Alliance group. We are a network of staff and students who work together, and support each other, with LGBTIQ and takatāpui matters. The group finds the conflation of people needing all gender bathrooms, and people needing accessible bathrooms, problematic and unnecessary. Rather, we advocate for the removal of unnecessarily gendered bathrooms.

The newest building on my university campus has some bathrooms designated gender-neutral. Unfortunately, the bathroom signs follow the usual conventions of two genders – a male stick figure and a female in a dress figure – plus a non-gendered body in a wheelchair (see Figures 4.1 and 4.2).

Figure 4.1 Gender 'neutral' bathroom sign at the University of Waikato.

Figure 4.2 Binary gender and 'accessible' toilet signs.

The women and men stick figure signs are an attempt to signal that the restroom is open to all genders, while the cubicle with a wheelchair symbol is a larger space for people of all genders, but specifically for those who have disabilities. These symbols and the gendered spaces they construct, are limiting in terms of gender diversity. There is always a disjuncture between actual bodies and the signs used to signify bodies on toilet doors. Restrooms for people with disabilities are characterised through the ubiquitous symbol of a genderless person in a wheelchair. It is the icon of access, 'yet it does not guarantee access, merely the *hope* of access' (Reeve 2014, 110). 'Ladies' and 'Gentlemen' signs, or their equivalents, create gendered architecture of exclusion turning the toilet into a micro space of hegemonic performances of masculinities and femininities (Ingraham 1992).

The attempt to create gender-neutral bathrooms caused some 'gender panic' in one of our new campus buildings. When I visited the restrooms someone had placed a temporary sign on each door (see the collage of images in Figure 4.3). The disability bathroom was re-gendered as female, while the other male / female bathroom was re-gendered as male.

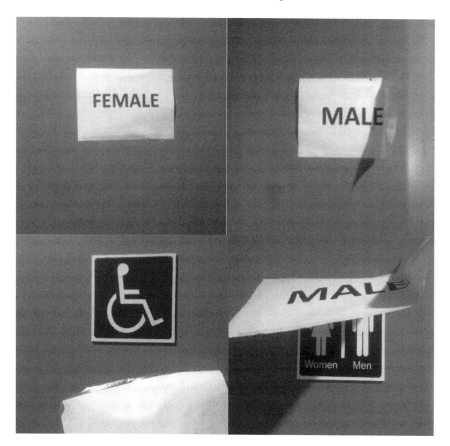

Figure 4.3 Re-gendering University of Waikato toilets.

For those who do not fit the normative gender binary of male or female, the temporary signs do a great deal of ontological damage. All of us in the University's Rainbow Alliance group (see www.waikato.ac.nz/student-life/student-experience/rainbow-alliance) insist that 1) there are many bathrooms that are unnecessarily gendered, and 2) there must be 'all genders' bathroom options for those who do not fit the narrow binary. Gender-neutral bathrooms are places away from cissexist surveillance and a way to validate diverse gender identities.

When I spoke to Sophie, who is intersex, aged 40–44, and Pākehā and works in a large government department, she said this about all-gender toilets:

> In my view, and I'm not like an architect, but I see it as, if we've got cubicles which have doors on them there's no reason why we can't have a row of cubicles with the wash basins. The whole area being unisexed. The cubicles are closed off anyway, and the whole thing is unisexed because each cubicle is closed off. So that means that women wouldn't have long lines, so to speak, because whenever there's a toilet available, there's a toilet available and they're not having to wait ... Washing your hands or even if someone wants to put on lipstick on, it's not going to be a problem because you're not naked when you're doing that stuff. And then people bring up 'well what about predation?' Well when anyone can pop in – whether it be male or female – there's probably less chance of that happening.

Sophie said that she uses the disability bathrooms at her workplace:

> I use the paraplegic toilets ... as much as possible I try to use them. But, an interesting situation regarding toilets is when I went to a ball, several years ago now, and I had to go toilet at that time and I said to my friend Cynthia 'I don't want to go [to the toilet]' 'cause I, for the first time ever, I'd worn a nice dress. Like I'd dressed up glam and I felt really good about it but suddenly there was this predicament and you go in the male one and it will just freak you out. Going to the female one, I was scared of the reaction it was like 'what are you doing here?' I think I probably, I can't remember now, I think she [Cynthia] might have come with me, I'm not sure. This was before my partner had come over and um, yeah but that was the, like the rest of the night was enjoyable but that moment was, it was freaky, it was scary. I'd hate to think about going to a nightclub ... whereas at least during the daytime, here, at least people are semi predictable because they're not drunk or anything. You add alcohol into the mix ... But even here you're still nervous if you have to go a women's bathroom, you're hoping there's no one in there and try to scoot in and out as quick as anything which is not good for your body system.

Gendered bathrooms, as experienced by Sophie, create unsafe and uncertain situations. Dominant cissexist and heterosexist attitudes lie behind the current obsession with gender-segregated facilities. The concept of 'safety' is worth exploring further here. Non-trans people invested in heteronormativity use dominant safety narratives to argue for gender-segregated toilets. The argument goes that women are likely to be assaulted by men therefore gender-exclusionary space is justified for women's protection. Real accounts of violence must not be diminished, yet, it is important to critically examine the role that hetero and cissexist discourses play in producing types of gender. The binary of aggression and victim underpins these predatory discourses with men constructed as out-of-control and predatory, while women are constructed as passive victims. Cavanagh (2010) shows that safety in bathrooms is far more complicated with dominant narratives of danger gendered in ways that reproduce dangers. Further, she notes another problem with heterosexist and cissexual safety: 'too often, trans men and trans women, butches, genderqueers, and those who are gay, lesbian, or bisexual are believed to be predatory, while violence against LGBTI people is rendered invisible and goes unreported' (Cavanagh 2010, 75). This is the case for Sally, a trans woman in her 70s, and NZ European, who told me about a sexual assault that happened when she went to use the bathroom during a party at a friend's house. The incident upsets the common idea that home bathrooms are safe spaces and public bathrooms are not.

> So I've had the ups and downs that women have. I had the worst incident when I was at a party. There were about 20 of us in the room, and there was a guy there with his partner. They were people I know and his name is Raymond. And his wife Carol has had her breasts removed because of breast cancer. So anyway and I couldn't tell this story – I don't cry now, I don't cry, but it really, really affected me at the time [Sally is upset as she tells the story]. And so we are having drinks and that and he said some debacle thing to Carol – 'why not get dressed like Sally? Look at her, look at her, look, dressed like that, be good to have breasts like Sally'. Which I found embarrassing. So anyway I go out to the toilet, and I go to the toilet and when I came out he's standing you know, right outside the toilet door in the passage way. And he's standing there and I say 'there you go Raymond' [gesturing that the toilet is free] and he put his arm around me and grabbed this breast here and, and I felt totally, totally powerless. That's why I start crying about it now, which I'm not going to do today [but is emotional telling the story]. So I go back inside again thinking, you know, I was violated, I felt um, I couldn't tell anybody, um so I went away I got a drink and sat in the corner, actually, and another woman came up and said 'what's wrong' and I said 'nothing's wrong'. I felt totally violated. And he came back into the room and was again as large as life, laughing and joking and I'm thinking you bloody asshole. And that taught me what women have had to put

up with for a long time and it taught me something else about myself that I am vulnerable and I can't stop that from happening. But it's just that feeling of, that feeling of vulnerability, yeah it was a really horrible feeling, not nice at all.

The retelling of this story was difficult for Sally. I have known Sally since I was a young teenager. As she told this story, the feeling of horror enveloped us as we sat together at the table. Sally did not report the assault to the police or tell anyone at the party about the assault. Sexual harassment and sexual assault – particularly in private homes – often go unreported (Lombardi et al. 2002). It is significant that she was sexually assaulted by a cisgender man at a private function in her friend's home. The public discourse about transgender bodies and bathrooms has centred on a type of gender panic about trans people, rather than cisgender people. In other words, the common conflation is that transgender people are sexual deviants and hence cisgender people panic that they may be at risk when sharing toilets (Westbrook and Schilt 2014). Sally is resolute, however, that this assault will not stop her from enjoying other night time entertainment spaces. She went on to tell me of another private function where she was 'dancing around with men' she didn't know particularly well, and where she felt accepted as a woman.

The policing of gender – through self-surveillance and from others – in bathrooms designed for heteronormative identities is evident in the stories from two participants – Cindy, who is transsexual, bisexual, in her late 70s, and New Zealand European, and Sarah, who is transgender, lesbian, in her early 50s, and New Zealand European (see also Johnston and Longhurst 2013; 2016). They are members of the Aotearoa New Zealand organisation Hamilton Pride (see www.hamiltonpride.co.nz) and regularly participate in trans activism and trans community events. Both Cindy and Sarah enjoy socialising in charter clubs (a type of bar and restaurant). The clubs attract an older clientele, who are mostly Pākehā (white New Zealanders), and working class. They feel accepted at the club, yet using their preferred bathroom is a problem. One day the manager said to Cindy: 'you can't use that bathroom'. Cindy was told that she had to stop using the 'ladies' and instead, 'use the handicap toilet'. Cindy responded that she is not 'handicapped' and therefore objected to using that toilet. She told the manager: 'I'm a female, I'm legally a female. I expect to have the right to use a ladies' toilet.'

I'm a female, I'm legally a female. I expect to have the right to use a ladies' toilet. We have now educated these people to the extent where we are able to comfortably use the ladies' toilets, but when I first started to go in them, I used to go in there and hope that there was nobody in there or wait until everybody was gone before I would come out again, not anymore.

Cindy is quick to respond to any complaints, when a manager of one of the clubs said to her: 'I don't know how to put this but we have had a complaint from someone about you using the female toilet.' Cindy retorted 'I don't know why. I am a female. You can't discriminate because I am legally a female.' She then showed the manager her birth certificate. To which the manager said: 'Oh, we'll have to change the database.' He went on to say, however, that he would prefer Cindy to use the bathroom down the back in order to 'save any hassle'. Cindy explained: 'I went down the back. I don't now. I use the main female toilet, too bad!' Sarah, too, had heard that some other clients had complained about her and Cindy using the women's bathroom:

> Yeah. I heard that, one of the barmaids told me that, that they still get complaints. They just say, 'even if you see the manager, he will tell you the same thing. Get a life. She is a female and that's it!' They know, the management knows, and I have been using the female toilets now for yonks [colloquial expression meaning a long time] and the manager never comes to tell me any different.

Both Cindy and Sarah took it upon themselves to 'educate these people' to the extent where they are able to comfortably use the women's toilets. Cindy reflects on people's changing attitudes:

> when I first started to go in them, I used to go in there and hope that there was nobody in there or wait until everybody was gone before I would come out again. Not anymore, but I did that to start with because I still felt self-conscious about it, that's just a learning curve I think, so there is a lot of hills to climb I think, hell of a lot.

In a different entertainment space – a casino – Sally, who is a Pākehā transgender woman in her 70s, told me a story about an incident involving casino management staff wanting her to use an 'out of the way' bathroom:

> I'm gambling away and one of the floor managers comes by and said 'excuse me, you see, we know who you are now. We weren't sure'. Because I'd been several times beforehand [dressed as a man]. He said 'we weren't sure who you were and we welcome you to the casino but we'd like you to use the toilets down stairs'. So I said, this was before I had my boobs done, so anyway I said, 'you know I didn't want to cause any problems' so I said 'oh yeah, yeah, yeah'. So we became best of friends. Then I spent a lot of my time there and I was up there one day and I was thinking 'wait on, you don't really look like a man, you should be able to use the [main] toilets'. So I went up to him and said 'Harry, I'm gonna use the women's toilets' and he said 'oh that's alright' [laughs]. I'm not sure what changed since then. I've done work, I've been up at the casino doing [food] demonstrations and stuff like that and we get on fine.

Sally, and also Cindy and Sarah, negotiate access to the women's toilet using a number of strategies. Initially, Sally didn't want to make cisgender people feel uncomfortable and she complied with the management order. Yet, she realised the inequity of this action and challenged it. For Cindy, her birth certificate is one of the resources she uses to show the fulfilment of all legal requirements. All women, however, are tentative with the establishment managers. They acquiesce and initially use the 'bathroom down the back' or the 'toilets down stairs'. Yet, in all cases, they challenge this instruction and decide to use the most convenient bathroom. Cissexist clients still complain, but in both of these entertainment establishments, the management teams support Sally's, Cindy's and Sarah's right to use the toilet they wish to.

Some participants told me that, due to safety concerns, not only do they police their own gender performance, they request others to comply with the rigid gender-segregated toilets. Lucy, a transgender woman in her 50s, shared a story about her relationship with her daughter:

> What I told my daughter is, I said 'you can call me daddy anywhere you want but just not in the women's bathroom please' [laughing]. So in the supermarket, you know, we'd be shopping and she'd go 'daddy' and I'd look around and everyone else were like 'huh'? And it became our little game, but she never would in the women's restrooms 'cause that would really upset things. Because we live in [name of place] and they're just not used to that sort of thing.

Lucy and her daughter play with gender in the aisle of the supermarket. In this public space they feel free to subvert heteronormative and cissexist space. Yet, the space of the toilet is not deemed safe to play with gender pronouns. In Foucauldian (1978) terms, when the body became an object of visual scrutiny, and hence subject to surveillance, the decision is to conform to gendered spatial norms. The majority of participants discussed public toilets. There are some, however, that shared stories about their private toilet at home.

Early nineteenth century settler homes – small cottages – in Aotearoa New Zealand did not contain bathrooms or inside toilets. Washing and toileting happened in separate facilities, usually 'out the back' of the cottage and connected to the outside laundry.

> In the last two decades of the nineteenth century, however, the bathroom entered the house – a very small room to begin with, just large enough for a tin bath and a basin of tin, enamel, or English porcelain on a wooden stand or cast iron brackets. Instead of a bucket to empty the bath, there was the novelty of a waste pipe which took the water to a soak hole outside.
>
> (Salmond 1986, 144)

The toilet, however, did not enter the house until much later. This was certainly the case where I lived – just north of Dunedin – for the first eight years of my life. The house was built at the turn of the century and the toilet was a standalone structure at the back of the garden. There was no plumbing, rather our excrement collected in a tin (the 'tippie tin' we called it) which, when full, was emptied (tipped) into a specially prepared hole in the ground away from the house and vegetable garden. Toilets began to appear as part of the laundry or near the back veranda around the 1890s (Salmond 1986).

People's home bathrooms may be places to try different ways to express and perform one's gender, in absolute privacy unless another family member finds out, as was the case for Lucy, as mentioned in the previous chapter on home. Lucy is aged in her 50s, is a transgender woman, lesbian, and white. She told me about the time her daughter found a 'hidden' wig in a bathroom. In the parent–child exchange, Lucy gently introduces the idea of gender diversity to her children and, at the same time, she needs them to honour her 'secret' until the time she is ready to go public and come out as a transgender woman. The space of her bathroom facilitated not just her own identity expression and management, but also a discussion with her children. In this instance, her private bathroom is a site of refuge, yet these are hard to find when in public spaces, as my last example in this chapter attests.

'Refuge restrooms' is the name of a web application for mobile phones designed to assist transgender, intersex and gender nonconforming people to find safe restrooms. The website states:

> When the Safe2Pee website passed out of functionality it left a hole in our hearts. REFUGE picks up the torch where Safe2Pee left off and makes the valuable resource available to those who find themselves in need of a place to pee safely once again. Users can search for restrooms by proximity to a search location, add new restroom listings, as well as comment and rate existing listings. We seek to create a community focused not only on finding existing safe restroom access but also looking forward and participating in restroom advocacy for transgender, intersex, and gender nonconforming folk.
>
> (Refuge Restrooms 2017)

As an open source web application 'Refuge restrooms' relies on users of the application to generate data. People who find and use gender-neutral and / or safe public bathrooms are encouraged to submit the location information to the web application, as stated on the web page 'if you know of a gender-neutral or safe restroom, please add it to our database!' (Refuge Restrooms 2017) and hence a map of restrooms is generated via the application user's location. The name 'refuge' is important. The makers of the application note: 'we firmly believe that everyone has the right to use the restroom in safety … Quite simply, we hope to provide a place of refuge in your time of need'. The name functions discursively to reverse the idea of bathrooms as risky spaces. The text on the webpage reminds us why we need the application:

> One of the biggest battlefields upon which the fight for transgender rights is taking place daily are restrooms. It seems that every other week a transgender child is made the center of a national news story because they used the restroom assigned to the gender they identify with. Obviously, we believe that every transgender person should have the right to use the restroom they want to. However, we also realize that despite legislative victories in recent years regarding restroom usage, many transgender individuals still face both verbal and physical harassment simply for using the restroom. Nobody should have to face that – and that is why we created REFUGE.

By way of example of how the app works, I typed in a U.S. city – Boston – as that was where I was preparing to travel to attend the 2017 Association of American Geographers (AAG) Annual Meeting. For gender variant and transgender geographers at this conference, the following safe bathrooms show on the Refuge restroom map (Figure 4.4).

The small number of 'safe' restrooms on this map is sobering. The database, however, is only as large as the users make it, hence the more listings, the more comprehensive and robust the resource will be. By touching an icon, full address information is available as well as comments on the 'usability' of the bathroom. For example, one person notes the following about a Starbucks Coffee bathroom: 'It says "no public restrooms" on the door, but I have been allowed to use it. Buying something and asking permission (or sneaking in when it's busy) might be a good idea.' At another Starbucks Coffee bathroom, the application gives more details about how to find the toilet. They also provide the following comments:

> You should probably buy a coffee or something, but I've used it when I really need to pee before catching a train (the bathrooms at Back Bay are disgusting, and also gendered) without buying anything and not had a problem.

The notes indicate that careful negotiations still need to take place, so these all-gender bathrooms are subject to the usual commercial transaction needed at Starbucks, that is, buying coffee.

Hostile reactions towards gender transgressions in bathrooms – public and private – bring into stark relief the performative and material consequences of binary gender norms. The toilet, bathroom, or public restroom, is one place that most participants said was a problem because of the rigidity of binary gender. Through the continual legal and social enforcement and maintenance of gender binary norms bodies are sexed / gendered in bathrooms. For bodies who do not 'gender conform' bathrooms may be hostile

Figure 4.4 Refuge restroom map, Boston, U.S.

spaces. Complying with cisgender norms is a significant stressor for trans-gender and genderqueer people. There are, however, several strategies that can be adopted such as avoiding some toilets, or by lobbying institutions to change toilets to 'all genders' spaces. The chapter began with examples of high school students lobbying their school principal and school governance board for inclusive bathroom spaces. They were successful. It's useful to re-member that the policing of bodies in bathrooms is based on a short history of what constitutes so-called 'civilised' bodies. Transgressing narrow ideas about gendered bodies and bathrooms may cause gender panic, yet such transgressions can affect positive social and architectural change.

5 Gender activism and alliances

Transforming place through protest

When Riki Anne Wilchins (1997a, 1997b), a transgender activist, was accused of transgressing the gender system, she retorted that it is the gender system that transgresses her experiences. This chapter considers the many challenges to the 'gender system', or as I would put it, challenges to 'gendered places and spaces'. Wilchins (2012, no page number) claims a political identity – for more than 30 years – that has been built on being visibly transgender from the day she: 'donned a Transsexual Menace NYC T-shirt and flew to the Brandon Teena murder trial in Falls City, Nebraska'. She recounts:

> Memorial vigils for slain transgender women, picketing HRC [Human Rights Campaign], books on gender theory and public fights with radical feminists, and being booted from the Michigan Womyn's Music Festival on multiple occasions for not being a 'born womyn' have made me who I am — inextricably intertwined with being publicly and very much a visible transsexual.
>
> (Wilchins 2012, no page number)

Transgender, intersex, feminist and queer activists and scholars concerned with social belonging, equity, human rights, civic duties, and gendered and sexed identities often engage in activism through organisations and / or individual action. Gender variant activism and activists' embodied emotions can be powerful forces for positive social change and challenge cisgender heteronormative places and spaces. They may also, however, reinforce identity hierarchies within and beyond activist spaces (Johnston 2017a).

If home, and familial relations within home spaces, are strained and difficult because of a person's gender, gender variant folk often seek support and assistance from activist communities and support groups committed to gender and / or sexual diversity and equality. The notion of community as a group of people sharing particular values, religion, ethnicity, sexuality or some other characteristic in common dates back to the Latin term *communitatem* meaning fellowship (Valentine 2001). Geographers have added a spatial dimension to discussions of community by drawing attention to the different scales at which communities are constructed. In this chapter I tease out issues surrounding activist communities in relation to gender variance.

It is generally agreed that transgender activism started to gain pace in the late 1990s in order to counter the stigma, marginalisation and discrimination felt by trans and gender variant people (Stryker 2008). Activist organisations helped bring to the fore the diversity and important differences among gender variant people. Research on transgender activism highlights: 'activists' definitions of *transgender* and the identities covered by this umbrella to inform an analysis of how different understandings of *transgender frame* activists' efforts for social change' (Davidson 2007, 60, italics in original). This research highlights the ways in which

> transsexual separatists, intersex activists, and genderqueer youth to trans-gender activists, gender rights advocates, and others organizing within the category *transgender* [alongside the politics of] inclusion and exclusion in terms of assimilation, social privilege, activist strategies, rights claims and policy changes, and the visions of social change forwarded by trans activists.
>
> (Davidson 2007, 60, italics in original)

I have worked alongside, and been an advocate for, a variety of activist community and non-governmental organisations (NGO) groups in Aotearoa, for example: Hamilton Pride Incorporated; Rainbow Youth; New Zealand AIDS Foundation; INA: Māori, Indigenous and South Pacific HIV Foundation; Te Rākei Whakaehu (a support organisation with a Māori structure for transgender people). Some of my research participants talked with me about their activist involvement with these and other organisations.

In what follows I first draw on examples of gender variant activism and advocacy through LGBTIQ Pride Aotearoa New Zealand organisations, such as the Auckland Pride Festival. Two Auckland Pride Festival Parades – held in 2015 and 2016 – shed light on the intersecting tensions of gendered, sexed, raced and class political action. Chen Misgav (2015, 1209) asks: 'what might be defined as spatial activism and how can it be implemented in a local-urban scale of LGBT politics?' I use this framework to make sense of local activist politics and paradoxes in a large metropolitan city. Second, this chapter rotates its lens outwards towards gender activism within international LGBTIQ organisations, and activism beyond the metropolis. Gender variant activism occurs in a myriad of ways in community group spaces, rural spaces and Māori spaces. Here, I pay attention to the potential of gender variant activism to radicalise conservative places and hegemonic institutional spaces.

Paradoxical activist spaces: (no) pride in gay parades

There is an established scholarship on pride activism and much of this focuses on LGBTIQ urban politics and community participation (Johnston

2005b, 2017a; Johnston and Waitt, 2015; Misgav 2015). As regularly claimed, in June 1969 several days of rioting sparked the beginning of what became known as a radical gay liberation movement (Weeks 2015). The Stonewall Inn of Christopher Street in New York City was the site where 'queens, queers and trans people fought back against the police' (Weeks 2015, 45). The riots turned into an uprising, and in recognition of these Stonewall riots, LGBTIQ pride groups – in most western cities – organise annual festivals and parades to advocate for, and celebrate, gender and sexual diversity. Christopher Street is now a famous queer tourism site (Johnston 2005b). Rainbow flags line the street as a welcome symbol to LGBTIQ people. Yet, not all rainbow communities feel included in this space, which is dominated by 'White middle-class and not necessarily tolerant of LGBTQ YOC [youth of color]' (Irazábal and Huerta 2016, 720). Gender variant people of colour, young people, and those who cannot be tourists are less likely to be part of this rainbow space (McDermott 2011).

There is now a lively debate within LGBTIQ communities about the benefits of annual gay pride festival and parade. The main issue is whether gay pride parades are effective forms of activism, or simply bright, sparkly and branded forms of commercialised homonormativity (Bell and Binnie 2002; Duggan 2002). Some scholars have written about the commercialisation of gay pride festivals in large cities and the ways in which genderqueer and transgender bodies become part of city branding (Bell and Binnie 2002); leading to a type of cosmopolitan climate (Rushbrook 2002). Furthermore, when regional authorities promote gay pride tourism, this is often deemed an indicator of LGBTIQ human rights progress. Yet, this may also establish some places as being considered more 'advanced or civilised' than others (Chambers 2008). Rejecting the commercialisation of Auckland Pride Festival, Tammy, who is 32 years old, identifies as transgender and genderqueer, and is Pākehā, says they are:

> not part of [Auckland] Pride. I don't [go]. Auckland Pride feels entirely corporate to me so not really like ... There are two things about that Auckland Pride that I [in the past] really loved. The first [thing I loved] was not really officially affiliated with Auckland Pride, but it was an exhibition that one of our friends kind of organised. That was just whole lot of really cool stuff with a huge emphasis on Indigenous people of colour kind of art and then Ollie also ran a little market. So you get to sell your arts and crafts and zines. It was in St Kevin's Arcade on K Road and it was just like our local kind of crowd and local people and low key and felt kind of nice. So those things I liked but I didn't really feel welcomed or celebrated at Auckland Pride Festival, or ... I was not really that interested. So no, not my cup of tea.

For many gender diverse and gender expansive people, these parades and festivals may be playful, political, and paradoxical (Browne 2007b; Johnston

2005b). They are sites where pleasure is at the centre of events that are also about resisting, challenging and subverting discrimination, marginalisation, intolerance and prejudice. New forms of LGBTIQ activism come to the fore as well as tensions between partying, politics, and commercialisation (Browne 2007b). I spoke with Joel, who is genderqueer, pansexual, aged in their early 20s, and Pākehā, about their experiences in Auckland Pride Parade. Joel drove a truck and talked to me about their body, feelings, and family:

JOEL: I was in the Pride Parade. I was driving the float with my mum.
LYNDA: That's so cool to be with your mum.
JOEL: Yeah and I spoke with my dad half way through it. Well not in the actual parade just before it. Um, and walked around the different floats so, so that was really cool.
LYNDA: Oh fantastic. So it all went well for you? You had a great time?
JOEL: Yep, yep I drew myself a beard on and um yeah it was great. It was really cool and like my favourite part was just kind of like sitting in the car and looking in the rear view mirror and just seeing how happy everybody was. Like there are all these people and there are giant smiles on their faces and they're just being themselves. It was so cool.

Joel talked with me about the pleasure felt by being surrounded by gender and sexually diverse people. The parade was significant for Joel because they felt closer to their parents, as well as the young rainbow community.

Feminist and queer geographers argue that place matters to gender politics and activism. There are many ways in which diverse genders and sexualities are spatially policed, marginalised, and valorised. The construction of geographical knowledge has, and continues to be, challenged by geographers who question taken for granted norms and seek new 'rainbow' possibilities. So too do geographers who actively participate in gay pride politics and parties. At the heart of pride activism research is that place matters to the construction, performance, and politics of gay pride.

With the ever changing nature and politics of pride parade festivals as a contextual backdrop, I became interested in gendered and sexed bodies at gay pride parades in Aotearoa, Australia, and Europe (Johnston 2005b). I am not the only geographer to track the gendered and sexed body politics of gay pride events, see for example: Australasia (de Jong 2015; Johnston and Waitt 2015; Markwell and Waitt 2009); Canada (Podmore 2015); Europe (Binnie and Klesse 2011; Blidon 2009; Browne and Bakshi 2013); and, the Middle East (Hartal 2015).

I now turn to some contemporary examples of gay pride in Auckland Aotearoa New Zealand. In particular, I focus on transgender activism in the 2015 and 2016 Auckland pride parades. In both of these parades the No Pride In Prison (NPIP) activist group condemned the Auckland Pride Board for allowing New Zealand prison corrections officers and police to march alongside the queer community after long attempts at negotiations on the issue. The intersections of gender, sexuality and ethnicity on the streets of Auckland can be understood as creating paradoxical spaces of activism.

It's common place to see politicians in gay pride parades whether this is due to 'pink-washing' – that is, using a rainbow-friendly images and practices as a distraction from human rights violations – or a deep and personal concern for gender and sexed justice. Yet, it was a surprise when in February of 2016, the right-wing, conservative politician and Minister for Police and Corrections – Judith Collins – marched in the Auckland Pride Festival Parade. Times have changed for this politician. In 2004 she used her parliamentary status to vote against the Civil Union Bill (an act designed to extend the same rights of heterosexual married couples to de-facto couples of any gender). She dismissed the bill as being 'about gay marriage'. Nine years later in 2013 she voted in favour of the Marriage Amendment Act, which allows people of any gender to marry and now she is happy to walk in the Auckland Pride Parade, along with uniformed police and corrections officers.

The inclusion of uniformed New Zealand Police and Department of Corrections staff in Auckland's gay pride parade is controversial. Experiences of homophobia, transphobia and racism defines the relationship between some members of LGBTIQ communities and police and prison staff. There continues to be growing support for urgent action on transgender rights, especially the safety of transgender prisoners. Transgender people are still fighting for control of gender identity, safety and access to healthcare. The Department of Corrections have made some progress over recent years but prison injustices still occur. Rae Rosenberg and Natalie Oswin (2015) in their article 'Trans embodiment in carceral space: hypermasculinity and the US prison industrial complex' found that trans feminine prisoners endure harsh conditions, including extreme gender regulation and reported being sexually harassed, with many subjected to sexual assault due to their gender identity. It is well documented that carceral institutions disproportionately affect non-white and low-income populations (Davis 2011; Gilmore 2007; Rodriguez 2006; Shabazz 2009; Sudbury 2002, 2005). Furthermore, Elias Vitulli (2013, 112) argues that

> the US prison system is also built on and produces systems of gender normativity and heteropatriarchy ... [and] critical prison studies must, therefore, centrally engage with questions of gender and sexuality and do so intersectionally with its analyses of race and white supremacy.

Responding to Police and Corrections in Auckland's gay pride parade, newly formed activist group called No Pride in Prisons (NPIP) made the following comment:

> 'Police and Corrections officers have no place in any pride parade. Both of these institutions deal violence to marginalised groups, including

queer and trans people' says Sophie Morgan of No Pride In Prisons. 'It wasn't long ago that the police were systematically beating and jailing people for engaging in non-normative sexuality and gender practices ... It is not because the police have had a change of heart. The police are part of an oppressive institution which has no place being celebrated in a pride parade.'

(Bedwell 2016, no page number)

NPIP went on to accuse the police of pink-washing and highlighted the poor police and prison record of institutional racism, homophobia and transphobia.

The 2015 Auckland Pride Parade was deemed, by many, a sparkling success. The national newspaper – the *New Zealand Herald* newspaper – reported

The parade is seen as the high point of the Auckland Pride Festival and was the first year police staff were given permission to march in uniform ... The parade largely went off without a hitch. The only disruption to the parade was a vocal group of three who protested the police contingent.

(Manning 2015, no page number)

It was the NPIP group who interrupted the parade. The group's message is that there is no place in a pride parade for institutions that fail to adequately care for the LGBTIQ people they are tasked with caring for. NPIP are particularly concerned for trans women prisoners who are often housed in men's prisons, placing them at high risk of sexual violence. They are also concerned about the high incarceration rates of Māori. Māori make up 50 per cent of the prison population, yet are 15 per cent of the country's population.

The 2015 NPIP protest was small scale – there was a banner, a megaphone and a group of people who stepped into the parade because they had an important message. The interruption was not dealt with well. NPIP member Emilie Rākete was admitted to hospital after having her arm broken by a Pride security guard. Emilie later spoke of her assault as the 'price of addressing Pride Inc.'s hypocrisy'. Emilie is a transgender woman and Māori. Emilie wrote this about the event:

This year a number of our members non-violently impeded the Police Force as they marched in pride of place in our parade and I, the only visibly Māori and trans member of the group, was assaulted by a member of Pride Inc.'s security, who severely fractured my arm and hospitalised me for days. (For the record – it was a broken arm, not a hairline fracture.) This is the price of addressing Pride Inc.'s hypocrisy. This is what Pride Inc. values. Marginalised queers need to stay in the margins, or we will be forced there with violence. This is what Pride Inc.

has made our parade into – an opportunity not to challenge the powers which destroy us and celebrate our victories over them, but to regurgitate their PR and participate in their marketing.

(Rākete and NPIP 2015, no page number)

Emilie Rākete and NPIP are committed to radical change, transformation and critiquing the Auckland Pride Festival organisation for allowing homophobic and transphobic institutions to be part of the parade. NPIP are not willing to become mainstream when injustices occur to gender and sexual minorities in Aotearoa. NPIP's place of activism is not within the parade, but 'at the margins'. NPIP wrote a statement about the groups' motivations: 'We are a group of queer and trans activists who did not want to see the violence of the settler colonial state 'pinkwashed' by the inclusion of uniformed Police and Corrections officers' (Rākete and NPIP 2015, no page number). Rākete and NPIP (2015, no page number) said that the crowd reaction to the protest was:

negative and violent. No Pride in Prisons' members on the sidelines were shoved away from the barriers and sworn at. They were told to 'get your own parade'. This is our parade. This is our space. Or at least it is meant to be. The reality is that what should have been a queer-centred space [instead] prioritised violent, racist, homophobic and transmisogynistic institutions.

NPIP protested at the following 2016 Auckland Pride, after the decision to allow Corrections and the Police to again march in uniform. Auckland Pride Festival board co-chair said they made the decision to allow staff to join the parade after Corrections agreed to better support transgender prisoners (Radio New Zealand 2016a, no page number). The Board understood that the Department of Corrections had been identifying issues, areas of concern and ways to do things better for transgender prisoners. This was enough for the Auckland Pride Board to release media, stating: 'For us, what we wanted to see was the potential for a positive change and the commitment to do that' (Radio New Zealand 2016a, no page number).

As with 2015, in 2016 NPIP again interrupted the Auckland Pride Parade in protest. A NPIP spokesperson said that the Department of Corrections were being rewarded for making promises to improve transgender prisoners' safety, yet,

It is not enough to reward the department for making promises it has yet to fulfil. Ultimately, the department is unlikely to make good on these promises, as up until now it has denied that it has a problem. The fact of the matter is that prisons ... are violent, racist institutions that have no place in any pride parade.

(Radio New Zealand 2016a, no page number)

NPIP held a rally that began an hour or so prior to the Auckland Pride Parade. There were speeches and about '400 and 500 people moving along K [Karangahape] Road towards Ponsonby Rd peacefully' (Yeoman 2016, no page number). During the official Auckland Pride Parade, police held back about 50 people who were trying to disrupt the parade. Some of the 50 protesters succeeded in halting the parade and held up signs that said 'Why celebrate those who mistreat us?' (Radio New Zealand 2016b, no page number).

These protests bring to the fore Indigenous incarceration and the complicity of Pride organisations in the ongoing colonisation and gender conformity of Aotearoa. NPIP group and their supporters protested at the march itself, citing the failure of corrections to protect transgender inmates from violence and the systemic criminalisation and over-representation of Māori in the justice system. Transgender people are 'told by the law, state agencies, private discriminators, and our families that we are impossible people who cannot exist, cannot be seen, cannot be classified, and cannot fit anywhere' (Spade 2011, 41). Trans and queer scholars, as well as prison advocacy organisations (mostly in the U.S.), are critically examining the incarceration of transgender and gender variant people in prison. The prison industrial complex (PIC) is: 'probably the most significant perpetrator of violence against trans people' (Spade 2008, 6).

Protests against Aotearoa New Zealand's treatment of transgender prisoners gains a great deal of media attention, due to the activism of groups such as NPIP. Corrections claim that in February 2014 they amended regulations 'M.03.05 Transgender and intersex prisoner' (Department of Corrections 2017) to streamline requests by transgender people to be placed in a prison that aligns with their gender. Yet, this process may take months and in the meantime, transgender prisoners are put at risk of sexual assault.

The acting co-chair of the Auckland Pride Board, Lexie Matheson, reminded reporters that the Department of Corrections were allowed to march in the Auckland Pride Parade in February 2016 as a result of their promise to produce a 'Transgender Action Plan' (Rowse 2017, no page number). Due to the failure to produce a plan, the Department of Corrections were not allowed to march in the 2017 Auckland Pride Parade. Emilie Rākete, of NPIP, says that 'seeing no plan appear after eighteen months is 'disappointing" (Rowse 2017, no page number). Matheson has a 'commitment to bringing about change with both the NZ Police and the Department of Corrections' and, 'what we want is for Corrections' practice to mirror policies. The policies are OK, the practice much less so. I guess staff who step out of line in this area should be held accountable' (Rowse 2017, no page number).

As already noted, flying the rainbow flag at pride events is increasingly seen as homogenising, and counter to the recognition of the intersection of subjectivities based on more than gender and sexuality, but also class, ethnicity, indigeneity, disabilities, age and so on. Other pride events, such as the Trans Day of Action for Social and Economic Justice in New York City, chose a more critical path to address the marginalisation of 'queer and trans people of color, low income people, immigrants, and people with disabilities' (Spade 2011, 206). When radically queer activists brush up again 'mainstream' reformist type organisations, such as Stonewall in the UK or the National Gay and Lesbian Task Force in the U.S., gender and sexual normativities are challenged. What constitutes gay pride activism, itself, is fought over, and interconnected with alternative economies of community socialising as well as mainstream highly commercial gay scenes (Andrucki and Elder 2007). The diverse ways of doing activism means that lines are sometimes drawn between radical change and assimilation to norms.

I asked all of the participants in my gender variant geographies' research project if they protest, are involved in pride organisations, or are part of community activist groups. Vonnie, who is transgender, Māori, aged in her 60s, knew of the NPIP protest. She was not convinced that NPIP did the right thing. While she was not at the parade, Vonnie had heard about individuals who had tried to stop it. Our conversation went like this:

LYNDA: At the [2016] Auckland Pride Parade there were people that stopped the parade and protested for the rights of trans folk. And at the same time there was another alternative parade happening on K-Road, did you hear about that?

VONNIE: I heard about the ones that were trying to stop it.

LYNDA: Yeah, and there was a group of people who got together – on K-Road – too, you know, for trans solidarity and to build capacity and to protest about what was wrong with the community. I totally understand the reason behind the protest, which also stems back from the year before when Emilie's arm was broken.

VONNIE: And that's what all the protesting is about, not being involved?

LYNDA: And also about prisons, spaces of prisons that are inappropriate for trans people.

VONNIE: Yeah, I totally agree with the prison bit. They [transgender people] should not be isolated in prison, but treated differently … I thought there were a lot of trans in the parade. What are they [the protesters] actually wanting?

LYNDA: There was a beautiful fa'afāfine presence in the parade, but I think the protests were about the No Pride in Prisons movement and against the involvement of Corrections and Police in the parade.

VONNIE: I haven't, I don't have an opinion on them. I just thought they stopped a good parade. You know there's another way of going about it. If they want to be a part of it they've got to join in or be part of the committee to change it.

LYNDA: Change from the inside?

VONNIE: I think they need to go and be part of the committee or when they have a change of leadership or anything be part of that process, to get voted in or get their supporters to put them in and say they need more diversity.

This exchange with Vonnie speaks to the heart of pride politics and para-doxes. While Vonnie is empathetic towards the way transgender people are incarcerated and treated in prison – she has experience of being imprisoned – she sees the Auckland Parade as a space for celebration. I asked Vonnie about her experience in prison.

VONNIE: Yeah, I went to prison in [an Australian city] for outstanding fines but they sort of put us [transgender people] in a hospital. It was like the main [male] prison was in one area and we were over in another area, you know, segregated from um (pauses) from other prisoners. You know we were able to mix. It worked quite well I thought, although they had males and transgender people in the same prison. You slept in your own cell and everything.

While Vonnie felt she was treated fairly in prison, she is very aware of her transgender sisters who are not: 'I know they've had bad times ... I've heard a lot of sisters talk about it'. One of the difficulties for transgender women in prison is that they are often placed in 'solitary confinement' in male prisons. Rosenberg and Oswin (2015, 1276) report on their findings, noting that 22 out of 23 participants surveyed:

> reported that they had been placed there [solitary confinement] at some point during their incarceration, for time periods ranging from 14 days to indefinitely, and often multiple times. On average, the length of place-ment in solitary for participants was 1.8 years per placement. Many participants stated that they had been moved to solitary as a form of punishment by prison administration, as well as for safety concerns (al-though most research participants indicated that they were still within harm's reach while in solitary).

It was against this systematic injustice that Vonnie and I reflected on the good work of a mutual friend – who is gay, Māori, and a leader in our rain-bow communities – who worked for the Correction Service in Aotearoa.

LYNDA: I think well, you know, we do have some good people there looking out for transgender people, but you know it's not enough. It needs a systematic change to be able to cope with gender variant people who go to prison.
VONNIE: We've got to change the [Corrections] Ministry eh.

We are both aware of the need to change policies, as well as cultures, of cisgender heteronormative and patriarchal prison environments. Prisons, like many institutions, regulate gender and enforce particular gender roles. Added to this, trans people are at high risk of incarceration. Spade (2011, 12) reminds us that trans people face a 'set of barriers – both from bias and from the web of inconsistent administrative rules governing gender – that produce significant vulnerability'. When discriminated against at work, in educational spaces, and rejected by family, gender variant people are more likely to experience poverty, poor health and homelessness. Accessing social services, healthcare, and housing, for example, means also experiencing everyday transphobia. If people do not have official identification that matches their gender identity and embodied performance, they often 'face major obstacles in accessing public bathrooms, drug treatment programs, homeless shelters, domestic violence shelters, foster care group homes, and hospitals' (Spade 2011, 147). In these precarious places 'low-income and poor transgender people engage in criminalized means of making a living' (Sylvia Rivera Law Project 2007, 11).

Over many years I have joined with other community activists to address gender injustices, inequalities and discrimination. In the following section 'Genderqueering community groups: everyday activisms', research participants discuss their experiences of collaboration and community action in order to create gender variant inclusive spaces beyond the metropolis.

Genderqueering community groups: everyday activisms

In this last section of this chapter I consider gender activism in queer groups such as: Hamilton Pride (www.hamiltonpride.co.nz); Rainbow Youth (www.ry.org.nz), University student queer clubs; and Māori marae (meeting grounds for particular iwi / tribes). The first space, an organisation called Hamilton Pride, is where Cindy, who is transsexual, bisexual, in her late 70s, and New Zealand European, and Sarah, who is transgender, lesbian, in her early 50s, and New Zealand European, felt accepted and part of Hamilton's rainbow community. On one occasion they presented their digital stories about their gender transitioning.

CINDY: That really made it for us, and for me, for me to become part of Hamilton Pride because at that stage our Agender group wouldn't do a blimmin thing and they had been talking about doing things for years and nothing was done. We were pleased to be with Hamilton Pride, to be part of it.

SARAH: I found it really humbling because of the turnout ... I mean there may have been other instances but I certainly, I was quite overwhelmed by the turnout and the support; it was brilliant.

Sarah is surprised and delighted that so many people in the Hamilton rainbow community came to view the digital stories and listen to Sarah and Cindy talk about the process of making digital stories. Cindy is frustrated that a transgender group had been inactive for some time, hence the rainbow collective – Hamilton Pride – provided the vehicle for transgender activism. The presentation was in the Hamilton's only gay bar (Shine), which has since closed. I estimate that about 100 people attended the event. While pride organisations are increasingly seen as 'mainstream' (in other words, neoliberal, capitalist, homonormative, and hence not radically transforming spaces and places (Misgav 2015)), the Hamilton Pride organisation is an exception as it is made up of a collective of gender and sexually diverse people who seek to be inclusive. Other participants who are transgender or gender-queer expressed similar feelings of belonging in small, localised, community rainbow groups. Amelia, who is aged 18–24, NZ European, MtF, and pansexual, in particular, enjoys being herself at Rainbow Youth and UniQ:

AMELIA: Since like going to Rainbow Youth and stuff my pool of friends has grown and stuff and since like going to UniQ which is completely, which is a completely different atmosphere.

LYNDA: What's that like?

AMELIA: Uni is more like, I guess, the closest thing I'd describe it as, it's like going to a bar instead of going to a support group. It's much more social and the people there are just as nice ... I mean they have their regular coffee meeting yeah, every weekend, oh every Thursday sorry not every weekend and yeah I go along to that, it's pretty fun.

LYNDA: And Rainbow Youth? Gosh they do great work.

AMELIA: They really do, I mean, and the thing about UniQ is it is a great place to hang out and meet friends.

Amelia went onto say that despite best intentions, there are still problems being genderqueer in these spaces. She says:

> Some groups are a little bit ignorant about trans people and more specifically more genderqueer people. I mean it's ... I guess it's a lot harder being genderqueer because you wanna be referred to as neither gender and that's ... hard to get people to identify you as that. I mean I personally am just happier as female for just now so that's a bit easier you know. And once people get to know me then, they can see me as female. Cos currently the world's separated out into different boxes, like you've got two genders and that's it. Sure that's slowly changing but .... [trails off].

My interpretation of Amelia's experience is that she is not suggesting that being genderqueer is a binary opposite to cisgender, but rather in genderqueer spaces one finds both cisgender norms *yet also* resistance / subversion of these norms (Misgav 2015; Puar 2005). Genderqueer space, then, is not a binary opposite to cisgender space. The space contains and resists gender normativities.

The spaces of Rainbow Youth are important for Joel (aged in their early 20s, identifies as on the masculine side of genderqueer, and Pākehā) who discusses their experiences:

> I used to be in a group at Rainbow Youth and so that was my first sort of official coming out I think, cos yeah I had to make it official. Cos,

our group once, ah at the beginning we sort of did our notices and stuff like that and then I just mentioned to everybody that I was going to be transitioning from being female to male and that I would really like everybody to try to use male pronouns and I did a bit of an educational thing about what that means or what's appropriate to ask and if anybody has any questions and they're appropriate to come and ask me and that went really well, yeah. And after that I went and came out to my parents a few months later.

In this space, it is important to consider the interplay of genderqueerness, race and ethnicity needs attention. As noted above, Rainbow Youth is not a 'pure' queer space (there is no such thing). There are hegemonic expressions of gender and these may be resisted and subverted. Gender is, however, further queered when considering race and ethnic identities, as Kiran, who is aged in their early 20s, is Indian / Chinese, queer, and transmasculine, highlights. Attuned to the power of binary cisgender, Kiran is also attuned to the differences in lesbian and trans masculine subjectivities in LGBTIQ spaces youth groups:

KIRAN: The group is weird because there's a lot of white queers and queer spaces are like that. It's irritating because I instantly **reek** in those spaces as lesbian. Um, which is a pain enough in itself partly because I'm not [lesbian] and also partly because I'm not that responsive to any sexual advances whatsoever [emphasis in original].

Kiran goes onto to discuss their frustrations with other LGBTIQ pride spaces:

KIRAN: Some people are extremely pro pride, like extremely pro of being in the Pride Parade and the rest of us are going 'this is a bad idea, you know this is kind of gross'. And go on for hours listing all these things – half of it's hearsay not actually true – 'I hear that in Pakistan they do this', and we're like 'oh my god you have bought into all this pink washing, god bless your soul'. But Rainbow Youth is a space that's generally free of quite a lot of transphobia, because trans people are in there, but it's still a space that's very, very white. And it being very white it's kind of weird and uncomfortable for me in ways because it reminds me that I've allied myself with something that is so dissonant from my cultural identity, if that makes sense?
LYNDA: Yeah it does, definitely.
KIRAN: And the queer spaces at Uni are the same.
LYNDA: So whiteness dominates those spaces?
KIRAN: Yeah and if it's not white people, it's people who are not white but who are basically white for purposes of queerness. That's a weird thing to say but it's true. You get a lot of gay and lesbian people who have nothing from their culture to base their queerness on and so they've

adopted a white queerness that involves a white politics that is kind of self hating in a way. That's going to be really toxic at some point.

By focusing on the interrelation of race and other embodied identities – gender, sexuality, age – Kiran brings to the surface the hegemony that LGBTIQ issues are synonymous with whiteness, homonormativity, and cisgenderism (McCready 2004). Media often depict LGBTIQ communities through the white bodies of celebrities and characters (Camilleri 2012).

Yann has experience of not being accepted in LGBTIQ community spaces. Yann, who is Pākehā, aged in their 60s, intersex and trans, remarks:

> I've had my least acceptance in the LGBT community. I've had more acceptance in the so called straight community, than I have with LGBT. And I actually don't think trans and genderqueer, or intersex should be part of LGBT. It should be separated out. I don't think they are the same as lesbians, gays, and bisexuals ... I remember shifting to Wellington, and I was complaining to some LGBT people there who were sort of organising groups and things like that and saying 'where do I go to meet people and things like that'? I used to be a lesbian, I still partner with lesbians. You know. And he said 'oh well you can just mix with all the trans'. And it's like, 'you miss the point'. It is that gender isn't about sexuality. Sexuality and gender trying to mix in one organisation is impossible because even if you are just thinking about it in terms of partnering – you've got 'lesbians and lesbians together' great. 'Gays and gays together' great. 'Bisexual – bisexual' well they will go whomever, yeah, so that's great for them. But for trans people, trans people don't necessarily want to partner with trans even though they quite often do. I think that is as much to do with the fact that the LGBT community, the lesbians and the gays, and the lesbians in particular are the worst, at accepting gender difference ... They don't like trans guys, especially the feminists. Although, that is changing. I actually think there's been a big change in the last two or three years.

Yann is criticising LBGT community spaces for being dominated by sexuality concerns, as well as cisgender lesbians, gay men, and bisexual people. Some geographers call this a type of 'queer patriarchy' (Nast 2002). Yann voices the struggle for those who are further marginalised in LGBTIQ community groups (the 'queer unwanted' (Binnie 2004)) because of their gender. While this has been Yann's experience over many years, they acknowledge that the politics of these activist spaces are changing.

Rainbow groups – Hamilton Pride, Rainbow Youth, and university student queer groups – have been at the forefront of same-sex marriage activism. In Aotearoa, the Marriage Amendment Bill was passed in 2013, making it legal for people of any gender to marry (or stay married, post transition).

During the lead up to the vote for marriage amendment, people were politically divided, as can be seen in Joel's reflection on the issue:

JOEL: I've got, like I don't personally believe in marriage, as such, or it's not something I would ever really want to do in my life but it should be my option not somebody else's. It's like, yeah I don't like the idea of somebody saying you're not allowed to have this choice, it's my choice and I can say what I want. So I'm really glad it went through. I think it got a lot of support because it was a bit of a, like it had, it was the sexy thing to support, yeah and I think things like gender, and you get things like child poverty and that sort of stuff gets left out because it doesn't make a good campaign cos people don't want to think about it. Whereas, yeah, when you think about the gay couple down the road, they're really friendly, and you can actually associate with that a lot more when you live in Remuera [an expensive suburb of Auckland] or something like that. Whereas if it's something quite far removed from what you're doing and your everyday life you don't see it and you don't, you're not inclined to help or do anything. Um yeah and I think the, I guess the things I've thought about with the marriage bill ... I like to think that the momentum will keep going but I'm concerned that it might just die. Which is a real shame... yeah there were some cuts to um to trans health care and that was really hard. I got quite upset about that.

Joel astutely compares political campaigns, noting that 'gay marriage' is sexy, whereas child poverty is not. In Butler's (2004b, 30) words, some lives are deemed 'more grievable' than others leading to the unequal spread of vulnerability depending on socioeconomic status, race, gender, sexuality, nationality, age and ability (Johnston 2017b). There is also a need to recognise that activism does not end with gaining the right to marry. Joel voices their concern over the cuts to trans healthcare and the lack of public attention and concern.

The final space that I turn to in this chapter involves gender variant activism in Aotearoa's Indigenous spaces. Māori-led activism – that is activism by and for Māori – was a topic of discussion by participants who attend marae-based events. A significant national event – Hui Takatāpui – is held for Māori gender and sexually diverse people. The Māori word 'hui' means 'meeting'. Takatāpui is, as discussed in Chapter 1, a Māori term that has been adopted by many to signify gender and sexual diversity for Māori. The event 'Hui Takatāpui' is a biennial meeting organised by New Zealand AIDS Foundation by and for takatāpui. In 2016, Hui Takatāpui celebrated its 30th anniversary and this coincided with the 30th anniversary of the passing of Aotearoa's Homosexual Law Reform Act (Harris 2016). The first Hui Takatāpui was held in 1986 and it brought together Māori rainbow communities. The hui responded to the AIDS epidemic of the 1980s and the homophobia often experienced within Māori communities. Jordan Harris

(2016), the programme manager for community engagement / Kaiārahi in the New Zealand AIDS Foundation says: 'we have emerged from the darkness of oppression and from the efforts of the early brave survivors paving the way, to standing with hope and pride on the Marae'.

Grace, who is aged 30, Māori, MtF transsexual, talked with me about her involvement in Hui Takatāpui. I asked her if she was attending.

GRACE: Yeah. Definitely I am organising a few girls, like trying to rally them to come because they don't understand the true meaning of takatāpui. They think its means something else which they are not into, so that they don't even consider to come. I just said to them, well if you are serious about our way of life you should definitely come to the meeting to show support because you need numbers, at end of the day for support... If you are not there, then what's the point? There's no point and nothing would happen if you don't support it. Yeah I am fully for it because it's like [getting support for] the medical side of our category. Those are the things we all strive for... I like to go out and live my life normally like everybody else, which is what we should be entitled to.

LYNDA: Excellent. Exactly. And I think the hui, from what I have heard from previous ones, that they are very supportive spaces for stories, for connecting, for celebrating, really?

GRACE: Yeah this is my first time ever coming to one because of my straight life and everything I was just disconnected. So I wasn't really... I was here but I wasn't. It was made hard for me with my issues and everything else. And now my life is about living to actually connect properly and get stable and to attend these things. I am happy I have got that far.

Grace is encouraging other takatāpui, particularly whakawāhine, to attend Hui Takatāpui. Whakawahine (which means 'like a woman') and 'tangata ira tane' (spirit of a man) are identities that come from traditional Māori concepts to denote those who were born with the spirit (wairua) of a gender different to what they were assigned at birth (Liddicoat 2008). Grace found Hui Takatāpui to be a place where she is accepted by other Māori, including whakawāhine like herself. She talks, however, of a reluctance by some to attend the hui, yet is convinced that it is a place for support beyond queer Pākehā dominated spaces. A participant in a study by Elizabeth Kerekere (2017, 173) expressed their frustration of Pākehā activist spaces, saying that Hui Takatāpui was 'completely different' and free from white-centric politics:

It's funny because when I was at the Takatāpui Hui, I was like, it's not like that [white-centric] here. We're all takatāpui, we're all here together and it's all sweet ... Also being at that Hui Takatāpui 2013, that day we rode horses and went down to the beach – that never happens in the Pākehā dominated world, it's like, 'we're here together. We're doing this thing together'.

Takatāpui activism also takes place on other marae, and not just necessarily during takatāpui-focused events. Vonnie talked with me about her role at a Māori iwi (tribal) sports event. The Waikato-Tainui Games is held every two years and reported as the 'largest tribal sports event attracting more than 20,000 tribal members who are all keen to represent their marae in the social head-to-head competition' (Waikato-Tainui Games 2016, no page number). Vonnie has represented the New Zealand AIDS Foundation there, providing information about being takatāpui.

VONNIE: Oh that was awesome ... at the Tainui games, it was awesome because we had a lot of people come up, 'Oh we should learn about this'. I said, 'Oh what a good idea!'... You know we get people coming up, looking interested and then we've got some guys come up looking, then walking past, then coming back to look at it again [laughs]. Go away, come back another hour or so, 'Excuse me'. So not able to talk about the subject, you know, just want them to feel comfortable. I just ask them where they're from and all that, how they are enjoying the games, although I want to ask about being gay or transgender, I thought, 'No, don't ask'. Let it come out in the conversation. You know we had a lot of enquiries.

LYNDA: Oh that's good and it's so important to be there at those events, to have a presence.

VONNIE: With us being there it's sort of opening up more, more so at the coronation[1] too, so all those Māori events and Tainui [iwi in the Waikato region], I try to make them aware that we're out there. And you see it all in the paepae [the meeting house benches where speakers sit]. Takatāpui standing up to do all the waiata, helping all the old kuia [Māori elder women]. But I think that needs to be changed. They're [takatāpui] are quite respected there but I think they're just getting used to us doing that part... But I think our kaumatua [Māori elder men] need to recognise how they are different, you know?

LYNDA: And takatāpui need a special place within that protocol?

VONNIE: Yes! Cause when I, when the Takatāpui Hui was going on I was going to nearly every hui in Tainui and getting up and speaking.

Vonnie explained the time and effort that she put in to make Hui Takatāpui an inclusive event. She went to all the marae in her region – following the poukai, which is an annual visit of all marae affiliated to the Kīngitanga (or Māori King movement).[2] So successful was Vonnie that now all the elders 'want to go to the next Takatāpui Hui'. Vonnie said 'I told them [other Māori in her region] that we [takatāpui] are part of your whānau'. Vonnie, Grace and other takatāpui have joined together to create a group called Te Rākei Whakaehu (2016, no page number), a:

Support organisation for Transgender people of all shapes and sizes, colours and any country. This group has a Māori structure but we welcome everyone. We definitely do not discriminate so please post positive things and anything to do with our transgender community. Kia ora.

Within regional geographies, beyond the metropolis, gender variant activism necessarily engages with multiple communities. The examples above show the politics of takatāpui and Māori-led activism and the way in which it is coming to inhabit marae and regional sporting spaces. For Māori, who claim their identities through links to ancestors and tribal geographies, connecting to place and whānau (family) is vitally important. What might be defined as spatial activism changes depending on place and communities.

There are also many activisms that this chapter hasn't explored. One, for example, is the formation of activist communities around gender variance in virtual spaces. During the past decade most of us have heard about, or even been engaged in, the phenomena of people using the internet to look up a variety of gender identities such as 'cross-dresser', 'transsexual', and 'intersex'. As one of the participants, Cindy, said to me: 'I typed in the word 'cross-dresser' and the whole world opened up to me'. There has been a recent and rapid uptake of video blogging, or vlogging, on YouTube by medically transitioning trans people (Raun 2012). Uploading vlogs about one's own gender transition is a personal and political way to create new gender variant geographies of digital trans activism. Eva Shapiro (2004) interviewed U.S transgender activists and found the internet helps trans people organise community and protests, find trans information, and challenge the pathologisation of trans and gender nonconforming people. Shapiro (2004, 172) argues the: 'internet has developed into more than a tactic or tool social movements employ; it has become a space – albeit a virtual one – within which organizing and activism can happen'. People of all gender variations use technologies such as blogs, forums, e-mails, social media, groups, games and (online and offline) events making online interactions and activist communities that form around the gender and sex nexus.

Gender variant activism is, undoubtedly, dependent on place and time. Back in 1969, transgender people fought police in protest over yet another raid of the Stonewall Bar, Christopher Street in Greenich Village, New York City. Today, Christopher Street may be lined with many rainbow flags indicating an inclusive LGBTIQ space, yet not all rainbow communities feel included in this space which, as Irazábal and Huerta (2016, 720) note, is dominated by 'White middle-class [people] and [is] not necessarily tolerant of LGBTQ YOC [youth of color]' (Irazábal and Huerta 2016, 720). This is a rainbow world that only some have won (McDermott 2011).

My final question for this chapter is: has gender variant activism become 'respectable' and perhaps devoid of radical disruptive change? Activists may fight for same-sex marriage so that people of any gender may stay legally married following gender transition, yet, there is plenty of evidence that state-sanctioned coupledom provides a narrow view of familial relationships, gender and sexual diversity (Wilkinson 2014). People committed to building inclusive spaces and places react to multiple and shifting hegemonies. A type of political agility is needed when responding to ongoing genderism, transphobia, sexism and misogyny.

Notes

1 The coronation, or koroneihana in te reo Māori, marks the time when the Kīngitanga movement appointed its first king in 1858. Koroneihana is a significant event for Turangawaewae Marae, in Ngaruawahia in the Waikato region, and attracts thousands of visitors from Aotearoa and beyond. Many iwi [tribes] from all parts of Aotearoa gather to support the Māori kingship. Koroneihana symbolises unity. See: https://koroneihana.com.
2 Before colonisation Māori tribes functioned independently, yet by 1850s Māori were faced with increasing numbers of British settlers, subsequent marginalisation, and pressure for their land. There was a belief that a pan-tribal movement could keep Māori land in Māori tribes, and provide another governing body for Māori. The Kīngitanga movement appointed its first king in 1858 (Papa and Meredith 2012).

6 'I find it a bit weird at work coming out'

(In)secure workplace geographies

In an interview with Joel, who is in their early 20s, genderqueer, Pākehā and pansexual, they talk at length about meeting new people in their work sales team:

> Every two weeks we have an induction meeting for the new ones that are coming into the workplace ... That's always a bit awkward because then you have to be coming out to the new people and I kind of, usually I leave it to the other people to tell them and I don't know if they always get the message ... I work with one other person very closely so we do the same sort of job for the entire sales team and um she's really good. She usually – we'll both walk into the induction meeting – she'll usually introduce me with male pronouns very loudly, so they kind of get the message um and then they look confused for a while and then she kind of reiterates it throughout the meeting and so a few of them get it, usually, um and then a few of them, I think, I assume they talk amongst themselves and figure it out. I like to think that's what happens. I don't like having to do more. I find it a bit weird at work coming out cos it's feels like you should be doing your work and stuff and yeah. I think there's about three or four people who really didn't get it and so I send them an email asking them to please use other pronouns. And then there's a few people who slip up every now and then but they're generally pretty good.

As a genderqueer person living in-between genders Joel told me about experiencing both feelings of inclusion **and** feelings of exclusion they had not previously felt. Joel's mixed feelings interfere with their desire to 'get on with work'. Joel is queerly in-between: occupying multiple positions as once, and different positions at different times, depending on how people read and react to Joel.

This chapter extends the fields of workplace geographies by discussing gender variant people's embodied experiences of, and feelings associated with, work. I draw on participants' stories to highlight experiences of seeking work, gender transitioning at work, 'coming out' and keeping work when one shifts from hiding one's trans identity to living it openly. I continue a theme that runs through this book, that is, gender is fluid and at times unknowable. Transgender, genderqueer, and other gender expansive people challenge gender normative workspaces. The notion of security – both ontological security and work security – influences how gender variant people feel about work. In a time currently characterised as precarious and anxious, feminist, queer and transgender social scientists are well placed to examine the (in)secure geographies of bodies, lives and labours (Johnston 2017b). Considering gender variant people's feelings reveals how power and privilege operate, and the possibilities of challenges to cisgender workplaces. The chapter provides accounts of gender variant people's embodied feelings of (dis)comfort and (not) belonging to illustrate the constitutive relationship between workplace, bodies and (in)security.

Before moving to participants' work experience, it's useful to review current research on embodied and work security. In a themed section of the journal *Social and Cultural Geography* (2014, 287) Chris Philo argues that emotional / affectual and psychoanalytic / therapeutic geographers have 'dispersed the human self' demonstrating that we are anything but whole:

> rather, fragmented into different, more-or-less connected regions … that may (or may not) impel the self to do things, to speak, to conduct itself or take actions in its life-world. In this vein, human selves become another order of 'strange attractors' which are themselves immaterial landscapes layering together different records of wider environmental, social, personal and (perhaps) sexual encounter, many if not all of which will serve to shape the self (or selves): helping to fashion its identity (or, likely, many identities); to provoke its desires and passions; to fuel its (dis)contents; to sustain or destroy its relationships; to enflame or dampen its anxieties; and so on.

Liz Bondi (2014) – in the same themed issue – extends understandings of the concepts of ontological security and insecurity. Using autobiography she presents a personal and subjective emotional geography of insecurity, with attention paid to the intense pressure experienced in conditions of material security and privilege.

Kristen Schilt (2010, 3), who conducted research on transmen and work, found:

> Not all transmen experience ostracizing reactions from employers and coworkers, particularly in retail occupations. Yet other transmen report more recognition and respect as men than when they worked as women – particularly white professionals who can physically 'pass' as men. These contrasting experiences show that gender boundaries can be policed (employers can encourage transmen to continue working as women and / or push them out of the workplace), or they can be flexible

(employers and coworkers can do interactional work to incorporate transmen as men at work).

Transgender people often are represented as adopting hyper-feminine or hyper-masculine gender identities post-transition, yet Schilt and Connell (2007) found that many also craft alternate femininities and masculinities. Despite these personal gender(queer) expressions, men and women co-workers continue to enlist their transitioning colleagues into rigid binary gender rituals. When this happens, political possibilities of making gender trouble in the workplace are limited. Transgender people wanting to avoid gender trouble may conform to binary gendered workplaces, thus avoiding the risk of losing one's job and friendly workplace relationships (Schilt and Connell 2007). People who may decide to live openly as a transgender person (due to the progress of trans rights activists (see Meyerowitz 2002)) may find that discrimination and marginalisation still exist in workplaces.

The chapter unfolds as follows. In the first section – Transactions and transitions: working bodies – the fleshy body is placed at the centre of analysis. Participants spoke about their personal decisions to conform to binary gender workspaces, as well as the ways in which their bodies disrupt and challenge normatively gendered workspaces. In the second section – Collegial workplaces: precarious power – the focus rests on body-to-body interactions with co-workers and clients, as I continue to flesh out the precarious feelings that some gender variant people have within their (cis)gendered workspaces.

Transactions and transitions: working bodies

Sally is self employed. She is in her early 70s, identifies as female and as a transgender woman, Pākehā, and has for two decades crafted and produced an artisan food product in Aotearoa. Sally lives in a small rural settlement, but travels to the nearest city where she sells her product. Sally is one of Aotearoa New Zealand's leaders of her product and the way in which it is crafted.

She sells her product to supermarkets and restaurants (throughout the country), and each weekend at a farmers' market. When Sally decided to medically transition she started with facial cosmetic changes and breast implants. In a three-hour-long interview, Sally explains:

> Now when I go to the market everyone thinks I'm the oldest butch lesbian in town I think since I've, I've got these [points to her breasts]. These [eyebrows] are tattooed on. This [points to eyeliner] is tattooed on. I wear makeup and I do eye pencil. My lipstick is semi-permanent and my boobs don't go away.

Sally's narrative is rich in content and highlights, among other things, her experiences of transitioning at work in a busy farmers' market in the centre of one of Aotearoa New Zealand's largest cities. As Sally medically transitioned from male to female she went from hiding her trans identity to living it openly.

The market in which Sally sells her product has over 65 different vendors and is one of the busiest farmers' markets in Australasia. It is in the centre of the city and attracts a great number of tourists, as well as locals. Sally has been a vendor since 2003 and is one of loudest sellers. She loves to draw a crowd to her stall, which is also staffed by one of her sons. She started to openly live as a woman in 2012. Here, she discusses her concerns about her changing body, and the way in which she self-polices her gendered behaviour.

> At the caravan at the farmers' market I arrive at 6 o'clock in the morning and I have to lift it [the awning] up every day. I've done it 600 times. I've been there 11 years now. But the interesting thing is since I've been taking the oestrogen my arms have got much weaker. My arms are still there, but there's no muscle any longer. I used to go to weight training and things like that until I changed, until I got my boobs actually. I was a bit scared of doing damage [to them] but now I can **just** lift it. I stand in front of the caravan psyching myself up to lift it. It's going to become more and more difficult. The slightest bit of frost on it will put me right off. The other day it rained and I was there, and it was about 2 weeks ago, maybe about 4 or 5 litres of water on top.

Concerned that she might damage her breast implants, Sally is reluctant to continue with weight training and now struggles when lifting the caravan awning. She attributes the oestrogen she takes to weakening her muscles. In spites of these challenges, Sally can no longer silence the gender she knows herself to be and has chosen to live with openness and integrity.

Also concerned with her changing body, Marlena – who is a truck driver, aged in her early 50s, transgender and bi female, and NZ European – spoke about her hormones:

> Well I, due to the hormones, and the fact that your testosterone is lower, and oestrogen is raised, you lose strength. So yes, some of the things I am doing I don't have the same strength that I use to have. But I'm not picking up or unloading the truck by hand. These days we have forklifts. The hardest job I do is pulling the straps tight, strapping down my load. Now that's not really too physical. I do notice that it is more physical now than it was before, but it's still not a big issue for now. When you talk about masculine industry – in the trucking industry – there's a lighter and a heavier aspect. Logging truck drivers, for instance, that's very masculine.

Our discussion highlights the importance of size, shape, and materialities of gender variant bodies at work. There are many gender variant bodies that do not fit a two-sex cisgender model. There are also people who exhibit gendered characteristics that do not align with the expected size and strength of their sexed body. In their article 'Making space for fat bodies? A critical account of "the obesogenic environment" Colls and Evans (2014) consider alternative theorisations of society–nature relations (among other suggestions). Here, I add to this field by considering transgender size and materialities in spaces that are normatively, and usually, dichotomously gendered. Transphobia may occur when one's body is deemed not the 'right' gender, but also the 'wrong size and shape', and therefore out of gendered place. While queer theorisation of sexed embodiment has rendered the man / woman dichotomy problematic and unstable, very little attention has been given to transgender, intersex, and gender variant embodied size, shape and materialities. There is, for gender variant people, a complex relationship between workspaces, body sizes, shapes and materialities.

In a news item entitled 'Big hands 'bad news' for beauty business' (Mann 2013) transsexual Stephanie Dixon says she is on the verge of quitting Christchurch because of constant questioning about her physique. The questioning is making her feel unwelcome in, as she puts it, 'conservative Christchurch' and she is considering a move to Aotearoa New Zealand's largest city, Auckland, where she may be accepted by a more diverse population where people are perhaps more open minded.

Stephanie claims to be Aoteaora New Zealand's only post-operative beauty therapist and she notes that the profession is her passion. She says that her business is failing because she has 'man hands' (Mann 2013). Stephanie has had gender reassignment surgery, identifies as a 'straight, single woman who drinks wine with her girlfriends and dotes on her two children'. She said things were going well in her business until clients found out she used to be a man. She has heard people ask clients 'do you know who she was' and 'what's it like with those hands?'. Stephanie goes on to say: 'There seems to be this sexual connotation, even though I'm a woman. They judge me as a guy. I'm being judged on the size of my hands.' Stefanie owns and works in her beauty salon and provides us with an example of a woman working in a feminised workplace, yet clients 'worked out' that she is transsexual (due to the size of her hands). Transphobia resulted in a downturn in clients for Stephanie.

In another physically demanding workplace – airports – I discussed with (a different) Steph her role as a security officer. Steph, who identifies as a transgender woman, is aged between 60 and 64, is white and was born in England and now lives in Aotearoa, is initially defiant when expressing her gender at work:

> I certainly found out that I could be who I wanted to be. It wasn't anything to be ashamed of. And I started going to work with female clothes on and make up on and I got in trouble at work.

Steph worked at an international airport as a security officer. As she started to transition, she felt confident and supported by co-workers, but not by her manger:

> I got a tonne of support, mostly from women, but from a lot of guys too. They thought I was incredibly brave. They thought it was a wonderful stance I was making, but they said 'you're going to lose your job over it'. ... I was getting more and more down about things, making mistakes, and they just tripped me up over things, watched me constantly, and had me in the office watching me. I told the boss I wanted to transition. She said 'I'll help you. You've got to get a name change, you've got to get a legal name change and then I'll open the door for you'. But she was just fobbing me off. Really the truth of it was that she wasn't interested in me at all. I was not really sure, not really, really sure about what I should do. I am going to lose my job and I am going to lose my family. I am going to lose my house. It was too much. Anyway it just happened that I resigned from my job. It's my job and I was stressed there. And of course now my wife wants a separation.

Steph's story is one of many told to me about the precariousness of employment when people decide to come out and live their lives as the gender they know themselves to be. Steph's insecurities with her boss and the stress of embarking on a medical gender transition meant that she decided to leave her job. In this excerpt she brings together a number of concerns: her job; her transition; and, her relationship with her wife.

Steph experiences new feelings that are 'particular to transness: anxiety, fear, hypervigilance. It can be dangerous to be a transsexual. It can be a lot of emotional work to navigate the cisgender world' (Nordmarken 2014, 38). As an airport security officer Steph had low income and a history of short-term casual employment contracts that could be ended or not renewed at any time.

Turning now to university workspaces, two participants (tutors) discuss their feelings about their bodies. When I asked Emily – who identifies as female, is aged in her early 40s, and of Asian ethnicity – 'what do you wear at work', she said:

> I like to wear dresses or skirts, sometimes jeans to work depending on whether I have meetings to go to. Very, very occasionally sometimes a jacket but I don't like it ... I observe others and try to learn from them but I feel frustrated in terms of how in this day and age people still have to, perhaps, unconsciously hide femininity. It's something that is regarded as problematic if you want to move up the totem pole. Of course there are 'successful' women, they are older, far older than me but they're always in black pant suits which is, I thought, 'mm okay' but that's not who I am. Students are more likely to challenge me, there

is that, but I still don't want to let any of the bias stop me from being who I am. On the other hand, I don't want to, okay I don't want to look, I don't know if it is fair to put it like this, 'ugly' or in all black, simply to gain some authority. Why is 'gender neutral' clothing always coded masculine? I want people to respect me for what I do, not how I dress. I want my freedom to express femininity. The corporate look is just not for me ... a lot of black. I don't wear crazy colours, well I like pastel blue dresses and stuff, but not all black ... I just didn't want to participate in their attack of femininity. What is so wrong with being a woman? Why do women have to dress in an androgynous way to gain respect? Why is girl empowerment all about emulating boys?

Emily analyses gender and power and refuses to participate in an 'attack of femininity'. She feels comfortable in feminine colours, rather than conforming to the way she perceives other successful women in the academy dress – in black pants and jackets. Emily's feelings of security – ontological and work security – and intimately connected. She takes a calculated risk by conforming with feminine ideals of embodiment; at the same time, she is cognisant that for women to be taken seriously in the academy, they must perform aspects of masculinity. This precarious position – Judith Butler (2004b, 2009) argues – exposes which bodies are at risk of not 'qualifying' as a recognisable subject. Butler (2009, ii) highlights the link between precarity and gender norms:

> since we know that those who do not live their genders in intelligible ways are a heightened risk for harassment and violence. Gender norms have everything to do with how and in what way we appear in public space; how and in what way the public and private are distinguished, and how that distinction is instrumentalised in the service of sexual politics; who will be criminalised on the basis of public appearance; who will fail to be protected by the law or, more specifically, by the police, on the street, or on the job, or in the home. Who will be stigmatised; who will be the object of fascination and consumer pleasure? Who will have medical benefits before the law? Whose intimate and kinship relations will, in fact, be recognised before the law? We know these questions from transgender activism, from feminism, from queer kinship politics, and also from the gay marriage movement and the issues raised by sex workers for public safety and economic enfranchisement.

Embodying gender norms in the workplace should make one less precarious, even less stigmatised, yet Emily is compelled to weigh up her options of dichotomous gender expressions. By opting to dress in a feminine way, she risks losing a position of knowledge authority usually associated with masculine performances. As Emily medically transitions, her body – she tells me – is under intense scrutiny by others in her workplace:

Early in my transition one of my colleagues – a female colleague who's a catholic and very conservative – said to me when I came out to her, 'oh it won't be any problem if you continue to come to work as you are dressed this week'. I was in androgynous mode by that point, my hair was pretty long but I wore women's jeans and tops. What she really meant was 'just tread lightly', yeah, don't cross dress. She is basically saying 'the only thing I can accept is if you continue to come to work as Harry, a guy'. I had to explain to her 'if I can do that I wouldn't have to come out to you. I wouldn't have to risk anything'. And secondly the assumption there is as long as you come dress professionally there wouldn't be any problem. I can definitely sense the irrational fear behind that advice is that she is really saying 'oh don't come in as a drag queen', or anything like that. Who would do that? I mean that's not what I'm about so I found that really annoying. Also from a very good girlfriend, close to both [partner's name] and I, she had the same confusion and worry about me being a 'drag queen.'

Emily went on to tell me that her colleague was relieved that Emily didn't turn up 'looking like a drag queen', and when she first saw her 'as Emily', remarked: 'oh that's great. You look fine just like the girl next door'. The continual reinforcement of cisgender feminine norms is something that Emily pushes back on. It has never been her intention to exaggerate femininity (as imagined and maligned by colleagues as the 'drag queen' look). Yet, she is also annoyed that her colleagues are relieved that she looks 'like the girl next door'.

Another participant who works in a government department – Sophie – spoke to me about being intersex, is aged in her early 40s, Pākehā, and running workshops. Sophie worries about student reaction to her embodiment, noting:

On days that I'm running workshops I try to wear tunics and leggings rather than dresses just so I don't freak some staff out. Not that I should have to be scared about that but I want to keep my job so. I know there are some that are probably born in conservative families that would freak out. Um and you just don't want complaints against your name so it's easier just to conform. And I think that's more of a sadness than anything else that you have to change just to make other people happy, not that there's anything wrong with wearing tunics. I like tunics too but you kind of limit what you're going to wear for certain days because of that.

The pressure to embody the 'knowing subject', that is a university professor, is felt by Sonny Nordmarken (2014, 37):

I wear a costume to make my class believe I am the instructor: professional khaki pants, dress shoes, a button-up shirt, an argyle sweater, and a tie. For the first two weeks of the semester, I am so nervous to perform 'instructor' that I do not sleep the night before I teach.

Nordmarken's (2014) discussion of 'feeling transgender in-betweenness' is an autoethnography of his everyday interactions living in a gender ambiguous body as he medically transitions from female to more masculine. He researches performativity, transgender microaggressions, and the emotion work transpeople do in everyday life. Building on Susan Stryker's (1994) use of 'monstrosity' as a tool of resistance and connection across difference Nordmarken's (2014, 37) goal is that 'people of all genders might see ourselves in each other, and that, together, we might work against gender injustice and social distance, and toward a deeper kind of intimacy and freedom for us all'.

Emily, Sophie and Nordmarken (2014) share their embodied feelings and performances about working in large institutions. Each encounters particular indignities and each draws on a variety of resources in which to navigate their workplace. Charting gender variant people's embodied feelings exposes other people's marginalising actions and reactions. Nordmarken (2014, 40) reminds us that oppression:

> is a form of collective trauma. It is inside all of us. We are not singular entities separate from each other—we all have multiple selves, and we all form a collective body. Oppression separates us all from parts of ourselves as well as from each other. And we need to work together with ourselves and each other to transform it.

Universities have long been constructed as places for 'rational thought', where supposedly emotion-free bodies conform to masculinist understandings of knowledge production. Transforming these spaces, ourselves and each other is a central task of another university professor, geographer Petra Doan (2010). Doan's (2010, 642) powerful account of her own embodied transitioning, documents her colleagues' and employer reactions when she came out at a university:

> As I entered the building I felt I was entering the eye of a hurricane, at the calm center of a turbulent storm of gendered expectations. As I walked down the hall I could hear conversation in front of me suddenly stop as all eyes turned to look at the latest 'freak show'. As I passed each office there was a moment of eerie quiet, followed by an uproar as the occupants began commenting on my appearance. Some people just stared, a few others told me how brave I was and one person told me that I looked 'just like a woman'. Another gave me a taste of what it means to be objectified by telling me proudly that I was his very first transsexual. These events helped me to realize that my presentation of gender was not just a personal statement, but a co-constructed event. I presented myself, and the academic world watched and passed judgment.

Doan's experience of transitioning at work highlights the objectification of her body, and at the same time it illustrates the ways in which these work-spaces are normatively cisgendered.

In another educational workspace – secondary schools – Mani, who is in-tersex, Pākehā, and aged in their early 60s, could never get permanent work, only in remote rural areas or in big cities schools where no one else wanted to work. At this point Mani 'ran away':

> I was really starting ... I mean I had questioned in my head for a long time, about gender identity. My questions were [also] around sexual orientation. So basically, I ran away. I headed off overseas .... I went to Kathmandu. I travelled over land all the way from Kathmandu to London. I went away for two years.

Mani shifted from teaching to nursing to working for people with disabili-ties with periods of unemployment. Eventually, Mani secured a job with the Ministry of Civil Defence, which they were excited about:

> I found my feet because, Lynda, in those days I had no emotional literacy, at all. So I found myself hanging out with guys exclusively. There were no women in civil defence then. They [Mani's colleagues] were probably as traumatised as I was because they were all Korean and Vietnam war vets. We worked completely in our heads. We followed operating proce-dures. We made our worlds safe by being highly organised. And I found myself in a place where I could survive. No unexpected emotion, in fact emotions were not welcomed in that era of the emergency services. My own private life was absolutely shit but no one knew about that. I was in a very violent abusive lesbian relationship. And so I had two worlds that didn't fit together. My professional world life thrived. While in my personal world, I was dying. That relationship and my job both came to an end at a very similar time.

Mani told me that through work, they used to embody 'male thinking' in a way that suppressed emotional responses when in a crisis. They eventually had counselling despite thinking 'counselling that was just shit for people who were weak' and it changed Mani's life. Mani retrained as a counsellor psychothera-pist and started doing their own work. This 'wandering career' – early training as an educator and work in civil defence with media training – meant Mani learnt to talk to people in a way that 'didn't frighten them', and this helped with Mani's camera work when producing documentaries on intersex issues.

Grace, who is in her early 30s, Māori, and MtF transsexual said she was 14 or 15 when she started sex work:

> I got into street working, and drugs at a young age, like I had to sup-port myself. I knew all the girls like myself and they are used to street

work. So they took me up to have a go and I sort of, you know, got into that way of making money. So that's how I got into it. And then it just became habit afterwards. And then drugs came along with that as well. That was part of my life but, yeah, it's not any more. But yeah, at least I can say that I survived. It was quite a hard life. And I saw and experienced a lot of things that people are traumatised by but I don't know if I am traumatised in any way. I just learnt to deal with that because I didn't have anybody to tell ... So I don't talk with many people about it because I don't think they will understand it. I think it's quite hard ... It's just hard for people to comprehend that some stuff like, you know guns and stuff, people kidnapping people, that side of is quite crazy. I don't say too much about it because a person can only handle so much ... I went through lots of bad stuff and nearly died quite a few times.

Being a sex worker has been one of the few employment options for transgender girls and women (Nadal et al. 2014) and Māori women remain disproportionately represented in the sex worker profession (NZPC 2013). Street sex work provided a space for Grace to express her gender identity, yet as she notes, this came with many dangers. High-profile Māori transgender women – Georgina Beyer (Beyer and Casey 1999) and Carmen Rupe – are takatāpui and whakawāhine icons in Aotearoa, and both published books about their lives, including their experiences as sex workers. During her time as an elected member of New Zealand Parliament, Beyer supported the passing, in June 2003, of the Prostitution Law Reform Bill. The Prostitution Law Reform Act decriminalised prostitution in Aotearoa and aimed to protect the health and safety of sex workers (Jordan 2005).

In supporting the Bill, Beyer spoke passionately in Parliament, relating her support for the Bill to her own background and work experience. For example, she stated that she was voting for the Bill for all the prostitutes she had known who had died before the age of 20, adding:

This Bill provides people like me at that time with some form of redress for the brutalisation that may happen in a situation when you are with a client and you have a knife pulled on you ... It would have been nice to have known instead of having to deal out justice myself to that person, I may have been able to approach ... the police in this case and say 'I was raped'.
(quoted in *The Dominion Post*, 26 June 2003,
see Jordan 2005)

Grace worked in the sex profession for many years, because, as she says:

GRACE: It took me a while to actually not go to the streets anymore and yeah it's taken a long time. I have been doing this since I was really young like 14, 15. It's a long time and I am 30 now. And I just stopped recently now, like four years ago, stopped working and yeah but I did work other jobs as well within my life time. But not many people, you know, know what it's like to be that young and working out in the streets where everybody is older than you. Because I have never met anybody who was my age until maybe couple of years later. Another girl just as young as me ... just younger than me came along. And then it was us and we stuck together all the time and did everything together because everybody else was older and they have been around for a while.

LYNDA: And hard to work on your own without some support of others?

GRACE: Yeah, I did have support from a few but street life is quite hard where everybody is focused on themselves ... and you can't trust anybody. You think you can trust that person but at the end you find out that you couldn't. And I experienced that most of my life, like people that I thought that were there for me weren't ... I went through period of depression for maybe a year. But I try to be happy in my life because I have lot more than those who won't have. I am lucky to be smiling [laughs]. Yeah, I am ok. I am happy, I guess.

Sex work is one of the most obvious examples of embodied work. The bodily contact is intimate, close, personal and sometimes violent. As Grace notes, she learnt to deal with trauma, but worked with many traumatised transgender sex workers. The street provides little protection for sex workers, where exchanges are usually in cash, in secret, and 'occurring in locations ranging from open spaces, waste ground, back streets and cars' (McDowell 2009, 101).

To return to Butler's (2004b, 2009a, 2009b, 2011) writing on precarious lives, the distribution of power across institutional, governmental, inter/disciplinary, and bodily contexts is uneven. Precarity – as a concept, condition and experience – is not a new topic for social scientists (Johnston 2017b). Geographers are considering precarity in relation to: migrant labour geographies (Lewis et al. 2016; Ettlinger 2007); population geographies (Tyner 2016); critical geographies (Waite 2009); squatting geographies (Datta 2012; Ferreri et al. 2017; Vasudevan 2015); geopolitics (Woon 2011, 2014); insecure body geographies (Philo 2014); and, sociology (Millar, 2017). There is value in thinking about precarity within and beyond the usual examinations of waged labour, and its central place within economic, social and moral geographies. By adopting and stretching concepts of precarity, precariousness, insecurity and vulnerability, feminist, queer and trans social scientists are able to examine how embodied genders matter when faced with the fragmentation of societal bonds, social and political governance, senses of entitlement and feelings of belonging. Experiences of precarity must be placed within 'the situation of relationality itself, insofar as our dependencies are vulnerabilities' (Puar 2012, 171).

Workplace geographies of precarity, precariousness and insecurities show which bodies 'do not qualify as recognizable, readable, or grievable' (Butler 2009b, xiii). In the following section, I consider, in more detail, precarious workplaces via body to body interactions, thus furthering understandings of relational gendered power in workplaces.

Collegial workplaces: precarious power

This chapter started with a quote from Joel. Joel identifies as genderqueer, and in Butler's terms, 'undoes gender' by being illegible (Butler 2004a). Joel occupies multiple positions at once. The sound of their voice is constructed as female, yet Joel (who has had top surgery and dresses in a masculine way) also experiences what white male privilege is.

Nordmarken (2014, 38) explains the feelings associated with being in-between:

> I feel the feelings I have habitually felt as a result of being positioned as female and treated as inferior. Yet, now people often position me as male in social interactions. In many of these moments, I experience a feeling of inclusion that I have not ever felt. Yet, at times, the femininity I continue to embody as a transmasculine being leads people to look at me funny.

Nordmarken, like Joel, experiences others in drastically different ways. People respond differently depending on how they read embodied gender. This complicated gender variant location is both Self and Other, or perhaps better expressed as Self-as-Other and Other-as-Self (Nordmarken 2014).

Thinking about 'in-between' bodies and collegial relationships in the workplace is useful as it gives deep insight into feelings of belonging (or not), as well as into gender variant people's subjectivities. In an interview with Yann, who is intersex, transmasculine, Pākehā, aged in their early 60s and pansexual, their embodiment is intimately tied to who and where they are, as well as what work they are involved in:

> Well actually at the moment I'm a support worker but I'm actually a textile designer by trade. I think that mainly, whether it is my homes or jobs and everything like that I've always been someone who's in transit to somewhere else. I'm a future person. I'm always a dreamer. You know, I'm always looking to um, what I can do next, yeah. I think I had a period there after India and after breaking up with my ex. I broke up just after I went over there after I discovered about my intersex stuff.

Yann said that since aged in their 30s, they had always been self-employed, hence work was usually always at home, or home was part of the workplace. In India, Yann worked in a factory warehouse, and also lived in one part of it. Work, home, indeed, all aspects of everyday life were co-constituted, leading Yann to reflect:

> I have in the past been tied more into **who I am in my work [rather] than who I am in my gender or in my personalities** ... There isn't a trade here [in Aotearoa] unless I set it up for myself. So um, I actually had to create an environment that was based around my gender or around my sexuality. Before that it has always been based around my work.
>
> (participant's emphasis)

Yann told me about a particular coping mechanism they used when living and working in India:

YANN: I use to have boxes and put things away. So that things looked tidy.
LYNDA: Is that, so? [pause]
YANN: It's not necessarily tidy cos when you open the boxes it might be full of fabrics or papers.
LYNDA: But you can contain it?
YANN: But you can contain it and put it away and you can, you can put it in places where it looks nice.
LYNDA: So why did you do that in India?
YANN: As a way with coping with the um, the chaos ... It was like the wild west there. You know you drove around in your jeep or on your bike. You know people would have guns on the back as, you know, you'd see people dead on the street, you'd see some horrific things.

In Yann's workplace in India, all aspects of life and identity collided. Feeling unable to control the workplace, per se, Yann resorted to placing work items in boxes in an attempt to gain control over the 'chaos' that was life, and work, in India. Recently diagnosed as intersex, and away from family and friend support networks, Yann said:

> I actually did most of my exploring of intersex and things like that while I was in, in India. Um, it was very isolated. They [work colleagues] didn't like me at all. They didn't like the fact that I was a lesbian and that I didn't fit in very well. My politics were very very different from the very classed society because of where I'd come from. I'm much more left wing.

Intercultural encounters at work were difficult for Yann, leading them to feel isolated because of ethnic, class, sexual, and gender differences.

Another participant, Jenny, told me about her workplace experiences. Jenny identifies as female (MtF), is aged 23, white, and at the time of our interview was not yet out:

> I worked at McDonalds briefly, [then] quit. I was unemployed for about a year from 2009–2010 before getting hired into my dad's company (who I'm not out to yet so my employment depends on how accepting he'll be).

Jenny had not experienced any discrimination in her workplace but was alert to the fact that it might happen: 'I will find out when I come out to my parents. The plan is to take hormones for as long as possible before it's impossible to hide, by which time the 'damage' will be done and I'll hopefully be able to convince my dad not to fire me.'

Jenny manages her health and well-being with this routine:

> On workdays I just get up, get ready, go to work, avoid interacting with people and quietly do my job so as not to be reminded of how everyone is perceiving me currently. I'm not out at work so I present as male and try not to think about it too much.

Jenny also spends 'a lot of time on the internet talking to people online at work, which helps with the isolation that comes with working in an environment full of out-of-touch real estate agents'.

When Sarah, who is transgender, lesbian, in her early 50s, and New Zealand European, talked to me about coming out at work, she said she could not wait to tell the other accommodation (motel) owners in her franchise:

> I brought my timing forward twice because I couldn't wait [to come out] and I wanted to do it earlier. Eventually, I had to go out and spend a whole day out in every motel in our group and tell them personally that I was coming out. I think that was a good thing for me. It gave me a bit more confidence and also I think it gave me more credibility in their eyes as well rather than just writing and saying what I was doing.

Sarah told me that the guests in her motel 'are fine' and that they show her more respect than when I was Bob. Sarah was the motel manager 'for three and a half years as Bob, and one year as Sarah'. One problem remains, however – when Sarah takes telephone calls from customers she is often mistaken for a man.

> I get 'yes sir, no sir'. Um 'excuse me, I'm a female', 'pardon?' [she laughs] and they get all embarrassed. It's so funny aye, I mean they get more embarrassed than I do.

Sarah makes light of this gender confusion. She is positive in her outlook, despite this daily misgendering. Butler (2009a, ii) usefully extends the link between precarity and gender norms, to ontological insecurities, for:

> those who do not live their genders in intelligible ways are at a height-ened risk for harassment and violence. Gender norms have everything to do with how and in what way we appear in public space; how and in what way the public and private are distinguished.

Joel, Jenny and Sarah do a great deal of emotional work in the attempt to secure their administrative jobs.

Sally, who I discuss in the first part of this chapter, says that in the farm-ers' market she is misgendered every day:

SALLY: So I'm standing there and every now and then someone will say 'ex-cuse me you didn't take your makeup off last night' or someone I don't know might say 'are you wearing a bra?' So you get this every day, every day someone says something.
LYNDA: Every day?
SALLY: Well yeah, these are out of towners, but that's alright, that's okay.

On the one hand, Sally explains this reaction is due to her in-between state of still looking male but also looking female. She wears makeup, doesn't always wear a wig at the market, and prefers to keep her hair short. She explained to me that the other market vendors know that she is transition-ing, but 'out of towners' – or customers new to the market – question her appearance.

In the ebb and flow of market life, Sally too feels the emotional weight of binary cisgendered norms where rigid categorisations of gender fail to include transgender people. Gendered bodies and spaces are subject to a regulatory regime that enforces the boundaries of normative gendered be-haviours. Bodies that destabilise male/female are disciplined (both individ-ually and externally) as people pass judgement on those who transgress the gender dichotomy.

Sally told me of an invitation to attend and speak at 'black tie' gala dinner at the Aotearoa artisan champion awards ceremony because of her role in the business.

> So I have this invitation well I want to talk but I want to go as Sally. So I email [the organisers] and say 'look if you want me to talk you need to know that Sally's going to talk'. I got all these emails back saying 'good on you', 'well done'. You know it was really touching, really, re-ally touching. So I arrive up there looking drop dead gorgeous. I go for a make-over and people I've known for, for years, from way back, we just embraced each other and cried in each other's arms. Why do we

> wait? And I know. It's something to do with being what you want to be. Looking drop dead gorgeous I speak and I get a standing ovation. And I'm thinking to myself, what is it about? Why have I waited so bloody long to do this and it's about, and I know what it is, it's about having the courage to do it. I have no illusions about it.

Sally found the courage to live as a woman. I suspect she received a standing ovation because she showed courage that evening in a cisgender male-dominated industry, and an industry in which she spent most of her time living as a man. I too am encouraged by her understanding of spaces that both constrain yet encourage embodied gender performances beyond cisgender normativities.

These stories of gender variant people's feelings of (dis)comfort and (not) belonging illustrate the constitutive relationship between workplace, colleagues and feelings of (in)security. Genders beyond dichotomies are a form of resistance and some people may feel powerful when their 'illegible' bodies undo gender (Butler 2004a). (In)secure workplace geographies, then, maybe characterised by the embodied emotions of those who do not qualify as 'legible', recognisable, or readable (again, in Butler's words). Attention to gender variant embodied emotions, helps undo the heteronormative 'tyranny of gender' (Doan 2010) so that new belongings are formed and felt, beyond cisgender normativities.

Dichotomously gendered work spaces influence the ways in which transgendered people express their gender. Yet, as with bodies, gendered spaces and places are not stable or fixed. The relationship between transgender and space is performative, dynamic and contingent upon place context. Yet, this relationship is also contingent on the heteronormative 'tyranny of gender' (Doan 2010).

In a joint interview Marlena and Steph, two transwomen, spoke about their work experiences. Marlena (aged in her early 50s) had been medically transitioning from male to female for two years. Steph (aged in her early 60s) had just had her first appointment with an endocrinologist prior to our interview. I asked Marlena 'how is it being a truck driver'?

> I'd like to work in a frock shop but I could never make the money I make working in a frock shop that I do make driving a truck. I'd love to sit in an office and be office staff, you know, but I'd only make about a third of what I earn driving this truck. Right now I need plenty of money to secure my mortgage etc. so I need to be doing the job I am doing. I need to be earning the money I'm earning. So I'm driving a truck. At first it was a bit frightening; it was a bit daunting getting a job. I had to convince an employer to take me on as trans.

Marlena's desire to change professions has not eventuated given the pay difference between what are considered feminised workplaces versus masculinised workplaces. She had been driving for several years before she decided to live fully as a woman. After transitioning, she found it difficult to get driving contracts. I asked Marlena about the reactions of other truck drivers.

> You get a bit of a 'stand offish' approach by some drivers, by some people, some forklift drivers, some dispatch people. They go 'I don't know about that one'. Once you get talking to people a couple of times they are fine. I have ended up with some very good friends out of it. One of the places I go to in Levin, the whole staff there are 'hey how are ya? Come on in. Come on in. How are you going?' It's like I'm treated so well by them.

Marlena, in general, provides a positive account of transgender acceptance at work. She is, however, mindful of the gendered nuances of her profession – truck driving – and she focuses on the places in which she feels accepted. Her ongoing concern is gaining future contracts. At this stage, she continues to drive freight and tanker trucks and continues to be appreciative of small town places, like Levin, where she is welcomed.

In another rural (Aotearoa New Zealand iconic) workplace – the sheep shearing sheds of the lower South Island– Grace, who is in her early 30s, Māori, and MtF transsexual, told me of her experience working as a 'wool handler' (also known as a 'sheep shearing hand', or 'roustabout / rouser') during sheep shearing season:

GRACE: Just like some areas where people wouldn't think that a tranny would go, like, like farm work, like shearing, because all of my friends are like saying 'did she really do that' because people don't imagine me [in that job].

LYNDA: Did you do rousing?

GRACE: Yeah, yeah. I did on and off for seven years. And there are couple of girls in the shearing sheds, older than me, but they came from Auckland. But it's quite a male-dominated area in shearing sheds.

LYNDA: Yeah. It's hard work from early morning to night. I did rousing for years.

GRACE: Yeah, they [friends] didn't think I'd get so dirty but I was younger, and I used to spend a bit of time like on farms and stuff with family, riding horses and that wasn't a problem.

Grace and I discussed the highs and lows of working in a shearing shed. Typically, the day starts early, bringing the sheep into the shed and shearers start their machines at 7 a.m. As sheep are shorn, wool handlers collect, sort, and bale the fleeces. The shed is loud with, on average, four or five shearing machines running, and music competes with the machines. The temperatures in the shed are extreme. It is cold at the start of the day and then the temperatures can climb to being unbearable in the uninsulated shed (Cotterill 2013). The social climate is also harsh for anyone who is not cisgender and heterosexual.

Grace continues to tell me what it was like for her to work in the shearing shed where her colleagues knew that she was a transsexual woman:

> It's quite hard to go to a place of work when people don't really understand you, a person like me. And they get a bit funny inside themselves … Oh when they didn't know and they were hitting on me and then when they found out they sort of freaked out about that and did not act the same towards me any more, which I experience everywhere. I don't know why people do that but I am, yeah, but sometimes they use their insecurities because they were attracted to me. They sort of blame me for their insecurity. And they get upset. But I am like, I haven't done anything. I said 'I haven't even done anything to you. It's like you were the one who was attracted to me'. And they ask me out and they find out. And they were all angry at me. I'm like, nah, that's not my fault.

We laugh together, both knowing that our laughter covers feelings of rejection and the pain of dealing with other people's anger. Grace's work colleagues – who changed frequently because of the nature of precarious short-term seasonal contracts – assumed she was assigned female at birth:

> They just saw and thought I was like that but didn't realise … otherwise at work I just keep to myself and maybe say hello here and there. It was sort of my defence mechanism not to get too close to the workers, um the male ones, because of what they may think of me if [voice trails off]. I just kept … I put a boundary up between us. Because they just can't get over the fact that I didn't look like the others girls they had there … Even though you may be nice looking you still get treated differently. And other girls were quite rough as guts, and if they [men] did anything they just get a belting from them. [Laughs] I was more timid and smaller than them [the other women].

'Rough as guts' is an Aotearoa and Australian colloquial expression and usually means someone, or something, lacks refinement or sophistication. I interpret Grace's assessment of her colleagues as being butch 'tough women', adding another layer of gender diversity to the shearing shed. Grace said that her colleagues 'had never met an actual proper transsexual' and despite the insecurities of many of her colleagues, some were 'very welcoming'. A few key supportive work colleagues became friends. Grace explains:

> that's what made me always want to go back to the [shearing] sheds. Like people would think it would be for the men, but it's actually not. It's actually because I enjoyed it and I liked the company. It was fun. And you can travel around. It's quite good, yeah, and physical. And

that's was good. I met quite a lot of good people and I am still friends with them now. I have seen them quite a bit over the years.

There's a great deal to learn about collegiality and precarious power from Grace's account of her shearing shed workplace. Insecure bodies – in this space – are not those who are transgender, transsexual, or butch. They are cisgender men who assume all bodies in the shearing shed are also cisgender. Some of Grace's colleagues were supportive of her as a transsexual woman and these friendships transformed the space of the shearing shed to one where Grace felt she belonged.

Other participants discussed the importance of trans colleagues in their workplaces. Vonnie, who is transgender, Māori, aged in her 60s and had sex work experience in Sydney – relied heavily on her work colleagues when faced with police harassment:

VONNIE: It was scary at the time, yeah. It was scary and exciting. I didn't stay at [name of car manufacturing workplace] and I was with a whole lot of other older trans and that's when I started transitioning with them and just went to work on the street. Oh we were looking over our shoulders every night.

LYNDA: How old were you then?

VONNIE: 17, 18. It didn't take me long to get onto the street. I was out and young and I thought nothing could stop me … My friend Stella was working on the street and anyway she came running into our flat and it was an old, old flat we were staying in, right on The [Kings] Cross, just opposite a tavern [bar] … And she comes running in [to the flat] and I'm standing in women's knickers. She runs through and jumps through the fence and I said, 'What's wrong?' The police were after her. There's me standing in my girly knickers going 'Ohh' and I had nothing up here [pointing to her chest] but it's just automatic to cover up here [chest], and put a hand down there [crotch]. So they arrested me 'cause they couldn't catch her or find her, yeah so I was in for it. But she came to bail me out and it was $100 bail and that was a lot of money back in those days.

Vonnie went on to tell me about systemic police corruption and how she and her colleagues had to pay the squad:

The police were corrupt over there. Cause before we could start you had to pay the consulting squad if you're known as a criminal … We used to put our $20 or $40 in our cigarette packet and pull it up and offer the police a smoke, then they'll take the money and then you're off, off you go but if you haven't got any money and you've just got cigarettes then they take the rest of them … But that still didn't work because they'll go and tell another squad to come and bust you.

In the absence of any organisational support (and police brutality) the informal support networks of trans sex workers was vital for Vonnie. The systemic transphobia experienced by Vonnie and her colleagues echoes that of many Māori transgender women who were continually harassed by police (NZPC 2013).

There is very little research within geography and the social sciences in general that discusses gender variant people's bodies, feelings and places of work. Yet, this is necessary in order to understand how power operates in workplaces and whether transgender people feel in / and or out of place. This chapter has canvassed the work (and out of work) experiences of a range of gender variant, intersex, takatāpui, and transgender people. Undoubtedly, gender variant people suffer discrimination and marginalisation at work (Whittle et al. 2007). Transitioning at work, applying for work, and keeping work were common topics discussed by many of the research participants, and these topics exist in published scholarship (Bender-Baird 2011), and media reports (Mann 2013). Workplace experiences illuminate the ways in which gender matters particularly when transgender people have work experiences pre and post transition. In other words, having worked on both sides of – and in-between – the binary, transgender people have unique experiences of gendered workplace practices.

These work experiences show how precarity – as it is embodied, felt, and encountered – is managed through self-care, collegial support, and finding spaces where one does belong. The research shows that the workplace may be both a challenging and accepting place for gender variant workers. What is clear is that there is an absence of organisational policies and practices in most participants' accounts of work experience.

7 Sexed up and gender fluid urban nightscapes

Inclusive spaces?

The bright lights of big cities are often hailed as promising windows into the imagined spaces of queer bars and clubs that welcome, and encourage, gender and sexual diversity. Yet, little is known as to whether these spaces are indeed unwelcoming for trans people. Browne and Lim's (2010) research in Brighton, 'the gay capital of the UK', found a city council website promoting gay Brighton's clubs, bars, shops, saunas, beaches and services with very little mention of the city's trans populations. Browne and Bakshi's (2013) research found that trans people's experiences of Brighton show that they are the most marginalised of the LGBT grouping. In other words, the benefits of a 'gay capital' impact lesbian, gay, bisexual, transgender, intersex and queer people differently. Across the Atlantic Ocean, Doan (2007) highlights that trans people are often excluded from LGBT spaces in the 'American city'.

Clubbing spaces, in particular, tend to be considered utopian and inclusive spaces where young people are free to express their identities (Chatterton and Hollands 2003; Malbon 1999) and provide opportunities for drinking, dancing, touching, (partial) undressing, using drugs or having sex in dark toilets (Jackson 2004). While many queer clubs and bars serve as important sites for constructing gender and sexual identities, they may also exclude the 'queer unwanted'. For example, elderly people may be excluded because of the number of commercial spaces in a youth-focused scene (Casey 2007). And it is not only 'old' people who may feel not welcome in a queer club. Other people considered 'unattractive' such as drab dykes (Browne 2007b), or non-white people (Caluya 2008) experience marginalisation in these spaces.

In an ethnographic case study of drag queens in Florida's Key West, Taylor and Rupp (2005) show the potentialities of nightclubs. Importantly, they highlight the ways in which drag queens unsettle the borders between man and woman, as well as gay and straight. Inspired by this work, I too conducted research on drag performance. The haptic geographies of drag queens (Johnston 2012) show the power of touch (body to body, body to place) to transform gender subjectivities and spaces. In another project, Chen Misgav and I (Misgav and Johnston 2014) consider transwomen's bodies, sweat and subjectivities in a Tel Aviv nightclub. Thinking through the body is also the topic of research into Detroit's 'Ballroom culture' (Bailey 2014). In another performance space – Chinese films – the politics of cross-dressing and nightclubs can also be found (He 2013), as well as films about transnationalism (Sandell 2010). Significantly, the research conducted in these 'club', dance and performance spaces highlights the ways in which gender is felt, celebrated, parodied, subverted *and* marginalized for trans and gender nonconforming people.

This chapter brings together research to consider sexed up and gender fluid nightscapes in cities. First, drag queen performances in queer bars in Aotearoa New Zealand are discussed. In these spaces gendered identities intermingle, sometimes producing diverse gender expressions, yet at other times, restricting and normalising gender. The second part of the chapter turns to the nightscape experiences of three transgender women – Cindy (who performs burlesque), Sarah and Sally – in 'straight' bars, clubs and other night time entertainment spaces. In all of these nightscapes gender diversity unravels gay and straight entertainment nightscape binaries.

Drag queening space down-under: touching moments

When people attend bars, clubs, and other night time entertainment spaces they expect to 'rub' up against other bodies. Some people may avoid bars for the very reason that bars tend to be intimate, dimly lit spaces where bodies are in close proximity to each other. I have a longstanding interest in drag queens in Aotearoa New Zealand. This interest started with my community activism and the organisation of annual gay pride festivals. Rather than just a spectacular sight and sound, I found that touch plays an important role in constructing gendered and sexualised subjectivities.

Bodily touching can reveal a great deal about people's emotional and affective relations with place. Sensuous experiences of touching, being touched and embodied feelings associated with touch can also provide insights into people's sexed and gendered subjectivities. When bodies touch it is the closest they can be, in the same place, and at the same time. As Kevin Hetherington (2003, 1933) notes: 'Touch produces a form of confirmation of the subject-world at the interface between the materiality of that world and the hand', although I would say 'that of the world and the body'. Touch may be everyday and mundane, yet it can also be highly political and hotly contested.

The word 'haptic' 'derives from the Greek *haptesthai* meaning of, or pertaining to touch' (*Oxford English Dictionary*, 1989). Haptic, therefore, means more than just the sensation of touch or the feelings associated with having one's skin touched and may be applied more widely to include embodied feelings and sensations. The notion of haptic makes it possible to think of touching beyond bodily binaries of inside / outside and crucially, as always constituted by place (see, for example, Morrison 2012; Obrador-Pons 2007; Porteous 1986; Rodaway 1994; Seamon 1980; Tuan 1974).

A haptic geographical approach – one that pays attention to the gendered and sexualised 'embodied experiences of touching and feeling, conjunctions of sensation and emotion' (Paterson 2009, 1) – can offer new ways of thinking about gender variant and queer people's emotional experiences and subjectivities. Some haptic geographies – bodies that touch places, places that touch bodies, and bodies that touch each other – prompt pleasure, pride and sometimes disdain, anger and pain, for drag queens and other club attendees in Aotearoa New Zealand. Furthermore, touch in this context may assert and subvert bodily and spatial boundaries associated with hetero / homo and masculine / feminine identities.

By way of a research example of 'down-under' gay night clubs, bars and dance parties Gilbert Caluya (2008) draws on autoethnographic research involving participant observation and informal interviews in order to focus on how racialised desires frame and structure spaces in Sydney's commercial queer scene. He is concerned with the ways that racialised gendered and sexual desires constitute spatial formations and practices that literally confine gay Asian males into ghettos within gay space. Racialised boundaries mean that Asian men – 'rice queens' – continually map Sydney's gay scene seeking places where they may find sexual partners and where they are less likely to experience racism. Caluya (2008) explores the ways that bodies that desire each other can connect without becoming reducible to race or normative constructions of gender.

It is important to underscore that what it means to be a drag queen has changed over time and space (Baker 1994). The *Oxford English Dictionary* (1989) notes that the term 'drag' was first used in 1870 and means 'Feminine attire worn by a man; also, a party or dance attended by men wearing feminine attire; hence *gen.*, clothes, clothing'. Leila Rupp and Verta Taylor (2003) chart the spatial history of drag in their book *Drag Queens at the 801 Cabaret*. They note that the term 'Queen' – originally 'quean' in the Queen's English, originally meant 'whore' and was used in the late seventeenth century England to 'refer to 'effeminate sodomitical men', sometimes called "mollies"' (Rupp and Taylor 2003, 180). In the U.S., the terms 'drag' and 'queen' did not come together until the 1930s (Beale 1989), with the first usage appearing in print in 1941 (Lighter 1994). A drag queen is a person, often a gay man but not always, who dresses and acts like a woman usually for the purpose of entertaining or performing. Like all gendered and sexualised identities there are numerous kinds of drag queens, or as they are sometimes known, drag artistes. Drag queens play with, highlight and subvert gender and sexuality by using their male bodies to exaggerate symbols of femininity. Drag shows tend to be 'camp' performances, that is, a type of recognised gay cultural practice that uses humour and audience banter to send up or parody heternormativity (Sontag 2002). Queens do not 'merely elaborate a directly feminine image; there is an articulation and refraction' (Besnier 1994, 308). Drag kings are often lesbians but not always, who act like men for the purpose of entertaining or performing.

The links between gender transgression and same-sex desire can be found in a variety of places and times. The dominant meanings of 'drag' – commonly used today in most western countries – have their origins in the mid-nineteenth century stage performances of female impersonators (Senelick 2000). The high point of public drag balls, in the 1920s, was in cities such as New York and Chicago, New Orleans and Berlin. Men 'might use the cover of masquerade to dress in women's clothing and dance with other men and where straight people came to gape' (Rupp and Taylor 2003, 183). Ironically, during the Second World War, drag became associated with soldiers, and the army provided scripts, music and even dress patterns (Bérubé 1990).

There is no doubting the ability of drag queens has encouraged feelings of pride, as noted in a tribute to 'old queens':

> They are, they have always been – whether famous or merely sitting next to you at the bar – the spotlights of gay life. They have always known exactly where to shine: on the very best, the very funniest, the most delicious aspects of being gay.
>
> (Rutledge 1999, 3)

Queens often have a strong emotional and political component to their performances. For example, the use of music may evoke complex emotional reactions from audiences when, through performance, drag queens deconstruct and destabilise sexual and gendered norms (Kaminski and Taylor 2008). There are drag queen professionals who have attained international reputations (RuPaul, for example), as well as people who may just try drag once. The movie *Priscilla Queen of the Desert* (Elliott 1994) – whose central characters, two drag queens and a transsexual, travel through Australia's outback to perform a cabaret gig in Alice Springs – is a useful example of drag down-under.

Nightclubs are important venues for finding community support. One participant – Vonnie, who is takatāpui, Māori, a transgender woman, aged in her late 50s, and works as an educator, visited many queer nightclubs to garner support for Hui Takatāpui, the biennial national event for Māori gender and sexually diverse people organised by New Zealand AIDS Foundation. As noted in Chapter 5, Hui Takatāpui is a national event for Māori gender and sexually diverse people. The event is, and needs to be, held on a marae (Māori meeting grounds). For some years, securing a marae was difficult, yet this did not deter Vonnie. She travelled from iwi to iwi, speaking to tribal leaders about the importance of the event. Once she secured the Māori King's consent, she said (emphatically): *no one stood in our way.* She then moved on to nightclubs, despite not being a regular nightclub attender:

VONNIE: So, once I got his [the King's] consent, *no one* stood in our way. They all came out in support in the end, you know? And it was my duty to go around [gathering support]. Oh and an old person, I was going out Friday night and Saturday night getting all the sisters in the nightclubs [laughs]. You know, I'd been there, done that. And walking up the street I saw my nephew's daughter outside a nightclub and she was yelling out to me from across the street and she's going, 'Nana, Nana!' And I was going, 'Oh shush!' [laughs] Oh, I am getting old! And I have to acknowledge her otherwise she'll keep on yelling out to me and, well she will just keep carrying on and they know I was getting embarrassed. And I was, 'Oh hello darling, how are you?'

LYNDA: You were great at rallying everyone together and pulling everyone together and making such a strong statement.

VONNIE: And I got into contact with some of our old girls, you know we all grew up together and we all lived together in Melbourne and Sydney. And that's how we managed ... we had one come back from Sydney and one come back from Melbourne to support the kaupapa [a set of principles and values]. We had all the Wellington sisters down there, you know, all doing the 'straight-lark' now but all came back out to support the kaupapa. So we had an awesome turn out by going up to Auckland by going nightclubbing and talking to everyone. Just letting them know what's happening and keeping them up to date.

This commitment was necessary in order to bring other takatāpui on board to support Hui Takatāpui. At this, and other pride events such as Miss Drag Queen Waikato, the Big Gay Out, and drag queen performances at Hamilton's gay bar 'Shine', I became interested in how takatāpui, transgender, intersex, genderqueer and gender nonconforming people feel in place and belong, yet at the same time, alienated, marginalised and excluded.

In Aotearoa there are several prominent drag queens who are iconic, indeed so iconic that five were employed as entertainers and crew for Air New Zealand's 'Pink Flight' and 'Flight of the Fairies'. Air New Zealand has attempted to reach gay, lesbian, bisexual and transgendered customers by becoming a major sponsor of the Sydney Gay and Lesbian Mardi Gras festival. Early in 2008 Air New Zealand launched its Pink Flight, a Boeing 777 that took passengers from San Francisco to Sydney for the annual event (see Air New Zealand's Pink Flight promotional video www. youtube.com/watch?v=bfhZ76SyVc0). This first ever North American gay-themed flight was decorated with pink décor (pink is a colour often used to reflect gay identity). The Boeing was decorated with a 50-metre pink feather boa and four-metre pink feathers were strategically placed on the windows to look like eyelashes. Drag queens and Air New Zealand's flight crew gave a night-club type cabaret performance at the terminal's departure gate. In addition, the Pink Flight offered onboard shows including drag shows, music, contests, screenings of gay-themed films and scheduled beauty sleep periods. In-flight attendants or 'hostesses' including Miss Ribena, Buckwheat, Tess

Tickle, Dallas Vixon, and Venus Mantrapp entertained and served guests (see Johnston and Longhurst 2010).

The iconic status of drag queens has reached regional cities in Aotearoa. Hamilton is a small city by international standards with a population of approximately 160,000. Celebrating gay pride in Hamilton often feels as though we occupy a position that in some ways is aligned with 'the centre' but in other ways is relegated to 'the margins' of the pride traditions in most western cities. Being out, proud and public has, Jon Binnie (2000) argues, increased visibility of 'mainstream' queer sexual cultures, yet it does not always feel this way in a small city surrounded by rural farming communities. It is perhaps surprising then, that a statue of Riff Raff, a cross-dressing character from the cult film and musical *The Rocky Horror Picture Show* (www.riffraffstatue.org), is located in downtown Hamilton (see Figure 7.1).

The film was written by Richard O'Brien, who grew up in Hamilton and plays Riff Raff, the butler. *The Rocky Horror Picture Show* is about Frank'n'Furter, 'a sweet transvestite from transsexual Transylvania'. The film adopts a queer perspective to the ways in which bodies become gendered and sexualised, included or excluded, depending upon place and time. Many Hamiltonians contested the proposal for a Riff Raff statue, yet overall support prevailed, and the statue was erected in 2004.

In a gay bar across the road from the Riff Raff statue, I have coordinated and supported events for and with gay men who do drag. Drag shows tend to open or close pride festivals, and performances raise funds and social awareness on World AIDS Day (1 December). There may also be performances at nightclubs. Drag queens here, and elsewhere, meet with a mixed reception even in supposedly queer movement times:

> Up against the assimilationist tendency of gay and lesbian activism, from the homophile movement of the 1950s to the present, drag queens have been an embarrassment ... however drag has played a central role in the construction of a public gay identity and has often been used as a political tactic in marches and demonstrations.
>
> (Taylor and Rupp 2005, 2118)

Drag queens work hard on their performances in order to get their message across. The use of comedy, music, satire and campy humour can be understood as a strategic form of political expression. Elizabeth Kaminski and Verta Taylor (2008) examine the ways in which music is deployed in four different ways to evoke emotional responses for their audiences at the 801 Cabaret in Florida, U.S. They note four different ways in which music is crucial to drag queen performance. These are: 'ritual' (which establishes and builds solidarity among queer communities); 'educational' (particularly for heterosexual members of the audience who are unaware of the experiences and grievances of gays, lesbians and transgender people); to

Figure 7.1 Riff Raff statue in Hamilton, Aotearoa.

'disidentify' with heterosexuality (by mocking, critiquing and challenging heteronormativity); and to facilitate 'interaction' between gay and non-gay members of the audience (Kaminski and Taylor 2008).

Audiences enjoy drag performances as entertainment, and at the same time, the performance is a site of contestation where gendered and sexed identities are negotiated and debated. The embodiment of gender and sex-uality is parodied and shows serve as arenas for the enactment or renegoti-ation of sexual subjectivities. The reactions to drag queens are varied, but what tends to be consistent is that many drag queens are touched and they touch others in ways that go beyond socially accepted 'norms'. I asked a manager of a gay bar about his views on drag queens and touch. Terry (who identifies as gay, male, Māori, and is aged in his early 30s) responded, at length, about the nature of drag as character and farce:

> I would say it is the exact equivalent of going to Disney Land and throwing your arms around, and having your photo taken with someone dressed in a costume there. Whereas if that person were not dressed up as a charac-ter you wouldn't obviously run up to them and throw your arms around them and so, it's just the same thing. I think people, when there's such an obvious barrier between an actual human being inside, then people feel like they can do that. They are not really interacting with the person it's just the character and the outfit they feel they are attracted to. And um, I personally don't have a problem with it. I know I have had to deal with drag queens in the past that have been a bit 'how dare you touch me, what makes you think you can do that kind of thing'. And I, as I say, I can see it totally from the other person's perspective, which is, well this is a *character*. I mean, drag, the origins of drag is farce, essentially, so I mean I get a bit uncomfortable with drag queens that take themselves a bit too seriously because I think they are losing touch with the origins of what it is they are doing, um, which is essentially mocking themselves and if people are going to um want to touch them and well that's just part of that process. I see it as a positive thing though. Because I think if people want to interact with you, that's just how they are expressing themselves.

For Terry, the notion that drag queens are 'characters' carries a lot of weight and contributes to the understanding that drag queens are 'sites of excess' that allow and invite bodily touching. Terry also makes a comparison be-tween drag and Disney characters. In Rupp and Taylor's (2003) study of drag queens in Key West Florida, U.S., they found queens who had engaged in drag as professional performers at Disney World, some of which had worked at Disney for 14 years. Hence there are drag queens that focus on the theatrical side of performance and create specific personas, as is the case with Russ, who identifies as gay, male, Pākehā, and is aged in his mid 20s, when I asked him about being 'Barb Wyred':

> Barb was basically made for me to go out, to be loud and noisy, and people to not to recognise me, as easily. Some people do, but not all of them. And to promote that we are a community, which is something I haven't been able to do, and I've always wanted to do and so she was made for that reason.

'Not being recognised' enables Russ to perform the excesses of drag in the character of Barb Wyred. He has a desire to contribute positively to LGBTIQ and takatāpui Hamilton communities by being a 'character' in order to challenge heteronormative attitudes. This is an important point about the excesses and politics of drag. Being out as Barb, however, can also mean that being touched isn't always welcomed. Touch may reassert boundaries between bodies and sexual subjectivities such as hetero/homo and masculine/feminine, yet it may also dissolve them.

In their study of drag queens in Key West Florida Rupp and Taylor (2003, 60) note that 'people do touch the drag queens all the time, in ways they never would anyone else, and the girls, in turn, grope them back. If a man they know comes by, they might fondle his crotch'. These experiences of touch also happen down-under. The complexities and context of these feelings within diverse communities are informed by media representations, such as the television programme *Takatāpui* (Māori Television, 30 June 2008). *Takatāpui* was probably the world's first Indigenous transgender, gay and lesbian television series. It first aired on Māori Television in 2004. The show provided insights into the lived experiences, and ancestral stories, of what it means to be Māori and gender and sexually diverse.

In a special show 'The world according to drag artistes' from an Aotearoa New Zealand gay, lesbian and transgender television magazine show, the narrator tells us: 'Often drag artists are grabbed, groped and felt up during an evening. The perpetrators are many and varied' (Māori Television, 30 June 2008). Some well-known Aotearoa drag artistes talk about their emotive responses to being touched. Felisha Peargetter/Whore from Gore reflects on one particular experience:

> I've had a woman come right up to me and grab my tits. I just turned on her and said 'did you enjoy that?' Right! I grabbed her tits. She went into shock and it's like, 'well, you felt comfortable grabbing my tits so I'm going to grab your tit'. She might say 'but yeah, yours are not real'. Yes [Felisha replies] but it still hurts, somewhere. Women are always the ones that pick you apart. You know, they sit there and give you the look over, assess you.
>
> (Māori Television, 30 June 2008)

This hostility in response to being touched was echoed by Dennis, who is gay, male, Pākehā and aged in his 60s, also known as Gloriousole, when I asked him if women (of any sexuality) were prone to touching drag queens, he confirmed that women do touch queens, 'especially they want to, they want to make sure that you're a fake'. Dennis continues: 'They grab them [breasts]. They will puncture them. And they will actually, I have come away with bruises from an event from people poking and shoving'.

Dennis went onto recall the times that Gloriousole had been prodded and poked at various events. It may happen in the context of play and fun, but it may also happen, as Dennis notes, out of a desire to confirm a particular embodied truth, that is, does she have real or fake breasts, and if they are fake, what do they feel like? And, 'what has she done with her penis'?

The popular expression 'seeing is believing, but feeling's the "truth"' reminds us that closely associated with intimacy, touch is used to confirm other sensuous information and to affirm contact between people and between people and their environment.

> Touching is closely associated with detailed evaluation of quality. Thus we feel the freshness of the fruit and the quality of cloth. If we are unsure whether to believe with our eyes, we will 'touch it and see', so verifying the truth like doubting Thomas.
>
> (Rodaway 1994, 149)

Touch here is not simply one body's interaction with another, but involves a tactile perception of bodily contours for both bodies. While touching engages the object (breasts, implants or bra inserts), it also invites an interrogation of gender binaries (Springgay 2003). Women that touch drag queens' breasts may do so in order to reflect on their own embodied subjectivity. In other words, touching may be gendered and confirm both normative and non-normative embodied 'realities'. Rupp and Taylor (2003, 60) note that one lesbian audience member of the 801 Cabaret in Key West grabbed a queen's breasts, because she 'wanted to feel what implants felt like'. This sensory experience of touch both dissolves and reasserts bodily boundaries.

Other eroticised bodily zones are also subject to sometimes unwanted touch. Russ notes that some people try to put their hands up Barb Wyred's skirt.

> You get the odd person who just has to try to put their hand up your skirt to find out if it [the penis] is real. It's like, excuse me please, that's personal space. I will actually kick you in the head, or something. I mean that's why I have such strong [hand held] fans and I will go through quite a few because I will actually hit someone with them. If I don't like you, and you're touching me, and I say no, then I will actually tap you on the head with my fan, you know.

Two iconic Aotearoa New Zealand drag queens, Buckwheat and DeeZa Star reflect on what it is like for them to be touched:

BUCKWHEAT: Depending on who it is and the way the context has come about, sometimes I can get a little bit [grimaces]. It's funny that they think they have licence to touch you, constantly.

> (from *Takatāpui*, Māori Television, 30 June 2008)

DEEZA STAR: Women, women love my arse. And I have threatened their lives and I have taken them outside and given them a twack. But no, they still persist. ... Women will break space with us because they think that it's okay because they are drunk and they're girls and it's not okay and they wonder why we get so offended. There is a bubble. If you break it we burst.

(from *Takatāpui*, Māori Television, 30 June 2008)

DeeZa Star's comments illustrate the way in which drag queens' embodied space is disrupted. 'Women will break space with us' is a reference to so-cially acceptable body-to-body boundaries (sometimes known as personal space) that are broken without consent. Perhaps women who feel that they can touch DeeZa Star's arse, or Gloriousole's breasts are responding to the way in which drag queens 'denaturalize' womanliness (Riviere 1986), or are the (more) perfect copy of femininity (as Butler 1990 might say). Drag queens are doing femininity in/with/from male bodies, and this mismatch exposes the sexed body as the base for gender as a fiction (Butler 1990). These exchanges also highlight they way in which being touched may re-assert boundaries between Self and Other, and straight and queer.

Dennis / Gloriousole offers another explanation of the sometimes antag-onistic feelings between drag queens and women:

The reason being is that if you're taking the piss out of women, women have every right to turn around and say 'hey!'. Women either enjoy the company or they hate the company [of drag queens]. Like, an exam-ple being if I go to a club of say heterosexual people, now, the men, some men, will find it wildly fascinating. Their wives will turn. Now, the wise woman will see their husband being titillated by that and realise the drag queen's actually got a claw into her husband and she will con-front him when she gets home, if she is clever.

Bodily boundaries of masculinity and femininity and heterosexuality and homosexuality are broken and re-asserted in this example of 'straight' men's attraction to drag queens. The excitement and fun that some straight men have with drag queens, as Dennis notes, may upset the wives of straight men. Heteronormative relationships are called into question when a 'hus-band' is having a fabulous time with a queen instead of his wife. Much of the scholarly work on drag queens has noted that their camp performances subvert the assumed naturalness of gender by sending-up notions of femi-ninity through highly decorated sparkly low-cut dresses, impossibly high heels, huge hair, and wearing make-up that exaggerates eyes and lips (Butler 1990; Rupp and Taylor 2003). In displays of further excess, they flirt wildly to applauding and cheering as well as lip-synching or singing 'gay anthems'. Such performances may redraw boundaries between heterosexual women and drag queens, yet, and often at the same time, boundaries between het-erosexual men and drag queens may blur. Subjectivities such as feminine and masculine, straight and gay tend to rely on distinction and separation to bring them into being, yet the role of touch may also 'resist, exaggerate, and destabilize distinctions and categories that mark and maintain bodies, signifying pleasure and desire as sites of insurgency' (Springgay 2003, no page number).

Touching between straight men and drag queens may happen, as Terry explains, because 'straight boys' will grab a drag queen's breast because they know that it's not real:

> We have our queens standing on the street and the straight boys all run up and they want their photo taken and they throw their arms around them. They might squeeze their, their, well, it's not even a breast, so um that's what gives them the licence to do it because they know it's not. So um, I don't know, I just find that sometimes drag queens take it a little bit too seriously.

It is possible that straight men treat queens as if they are cisgender straight women and touch them. As one of the Key West drag queens, David, noted:

> They'll touch the ass. They may run their fingers over the breast. It's the butt and the thighs and the legs. Running the hand up and down the legs and the knee and the thigh ... And they want to stroke your hair.
>
> (Rupp and Taylor 2003, 60–61)

A gay male audience member of a drag show said that 'when I was in drag, people were more comfortable with coming up to me and being touchy and feely and not meaning anything by it' (Rupp and Taylor 2003, 61). He thinks because women are 'allowed' to touch and be touched that this status extends to drag queens. These examples of touching highlight the volatility of binaries hetero/homo and masculine/feminine. This example also highlights that an appreciation of the sensuality of touch needs to also be accompanied by an awareness of the politics and complexities of gendered and sexualised subject positions.

Doing drag prompts a range of feelings, as Geoff (who is Māori, gay, male, in his late 30s, and also known as Sefina Soul Train) reflects on his involvement in Hamilton Pride 07:

GEOFF: I also think the Red Party was a fabulous event for community spirit, especially for what the Red Party was all about, to honour those who have gone because of HIV and AIDS.

LYNDA: It was great. And you looked amazing.

GEOFF: I was actually quite excited about doing drag because it had been so long, 2000 was the last time, oh no, 2004. I wore that same outfit but with a big huge head dress. I think the buzz for me was just to do drag and to get up on the stage because it's, there's, there's this fierce drag queen within me that just wants to get up there and just be fierce in fact.

Just be fierce then get off and go home and know that she's had fun and she hasn't hurt anyone you know and I don't think I did that night.

I asked Geoff if he had experienced being poked and prodded while in drag:

Yeah, um, yeah if they are straight women they just poking and prodding out of curiosity. But some gay women actually feel up drag queens in, ah, almost like a sexual way. But I don't know if it really is because they're lesbians and wouldn't normally be attracted to a drag queen unless you're really tranny drag like a hot woman and there are a couple of those ones around. Unfortunately, I wasn't one of them [laughs]! Not that night! I was full drag *queeeen* (emphasis in original).

Geoff emphasises that for Sefina Soul Train drag was exaggerated femininity, which is unmistakably drag and not transgender. Similarly, drag queens such as Tess Tickle, are not offended by being touched, nor mind if it is lesbians who are touching: 'Now I get a lot of the lesbian community [laughs] who love to pinch my arse or come onto me, which is cool with me, it doesn't matter to me' (Tess Tickle from *Takatāpui*, Māori Televsion, 30 June 2008).

Sefina Soul Train and Tess Tickle are very relaxed about being touched by lesbians suggesting that this embodied experience of drag queen does not map easily or neatly onto existing familiar gender and sexual subjectivities. Similarly, Rupp and Taylor (2003, 60) who identify as lesbians, said that 'Sushi' – one of the queens in their study – 'once put his hands on both of our pubic areas, announcing coyly, 'It's a service''. What these encounters between drag queens and lesbians tell us is that touch enables an intermingling of bodies, genders and sexual subjectivities. These intimacies point to a complex set of emotional intensities and sensory practices that may transcend sexualised subjectivities and desires usually associated with being drag queens and lesbians. Desire, as has been noted, 'is messy and does not obey the obligations of identity' (Lim 2007, 63).

To underscore the importance of place in the construction of genders and sexualities in nightclubs, one can look across the globe, to a nightclub in Tel Aviv, Israel (Misgav and Johnston 2014). My colleague, Chen Misgav, and I worked on a nightclub project together, and in doing so added to knowledge about the co-construction of genders, sexualities and nightclub spaces via bodily touching. Chen and I analysed the spaces of 'Haoman 17' – one of Tel Aviv's biggest clubs in the city. This club is spatially distinctive with transgender women (often called Coccinelles in Israel) and gay men occupying different spaces. Coccinelles wishing to maintain aspects of their femininity distance themselves from dancing, sweating, and drugged gay men. It is common for gay men – in this club and many others – to dance bare-chested. The Coccinelles, however, who carefully prepare their bodies for the nightclub, wear their best clothing, and try not to sweat or come into direct contact with sweat. The Coccinelles 'keep their distance' from sweating gay men's bodies, and also particular rooms in the nightclub. Sweat, in this nightclub, becomes defined as masculine and as both in and out of place depending on one's gender expression and identity (Misgav and Johnston 2014).

As the above shows, there is some research on the importance of entertainment nightscapes for drag queens, and rainbow LGBTIQ communities. There is less research, however, that speaks to the experiences of transgender people in 'straight' entertainment night time spaces (in other words, in entertainment spaces that are not specifically advertised or recognised as queer or LGBT friendly). I'm not suggesting that there is an easy binary between gender 'queer' and 'straight' night time entertainment spaces. As the following examples highlight, the boundaries are porous.

From burlesque shows to private parties

Aotearoa boasts having one of the world's oldest burlesque performers. In 2016 Hamilton-based Cindy – Miss Chevious Cinders – turned 80 and she celebrated by taking part in the Burlesque Baby show (Austin 2016). Cindy, who is transsexual, bisexual and New Zealand European, started burlesque five years earlier. In Chapter 3 I discuss her home spaces, particularly her lounge, which is a homage to burlesque. I asked her about how this started:

> It all start back somewhere around May 2011. And I was helping a lady, who you know. She was setting up her hall and decorating the hall. And the burlesque girls were there practising because they were going to perform that night, and they were doing a practice. I started doing the bumps and grinds to the music. And the head lady come up to me and she says 'oh we have classes here you know.' And I went home and thought about it and I emailed her and said 'Do you mind if a transsexual comes along? I am legally a female.' She wrote back and said 'No, by all means come along.' So I started and now look what happened. [Laughs]

Cindy immersed herself in burlesque practice and performance, and she is now known as 'New Zealand's Grand Dame of Burlesque' (Glamilton Burlesque Academy 2016). It started with small classes, then performing for audiences around Aotearoa New Zealand. Cindy also went to Melbourne to attend shows and to take classes. She said:

> I went to Melbourne to meet Catherine Delish who is one of the top Burlesque girls in America and she was out there and we spent a Saturday afternoon with her. It cost us $200 to go, but it was worth it. We met one of the biggest stars of burlesque and we were amazed and that got me going again. Since then, of course, I am performing in Hamilton

here with the Bombshell Burlesque Starlet Review. I have done two shows with them. And when I did a show with them I was one of the guest performers.

Cindy said that the highlight of her burlesque career, to date, was:

> being accepted to perform in the New Zealand Burlesque Festival in Palmerston North in October, on the first night which is October the 9th. That was quite a surprise. There were 150 applications to perform and I just thought well, I will give it a go. I had my doubts because all the others were pretty high up. And I got quite a shock when I got the email saying I was accepted.

Burlesque classes and performances are usually attended by cisgender women and, very occasionally, cisgender men. This does not deter Cindy, rather, she feels completely accepted at burlesque classes and performances, despite admitting to be initially shy in front of an audience. She says that the burlesque family is a very tight-knit community, 'no matter who we are, what we are, we are all family'. She elaborates:

> We all do the same things. We are all in the same sort of boat. And we like to perform in front of people. I never thought I would ever get up on the stage, I remember being at the Outgames in 2011 and Mani Mitchell had to hold me down. I was shaking like a leaf talking to 50 people. [Laughs] I was just shaking like a leaf, I had never done it before.

Here Cindy speaks about feeling nervous in front of queer crowds at the 2nd Asia Pacific Outgames Human Rights Conference – a sport and cultural event held in Aotearoa's capital city, Wellington. Cindy presented her digital story about her life, her transition from male to female, and her role as an advocate for other transgender people. She talks about this digital story online (see Lewis 2011). Cindy told me that she felt nervous because of the seriousness of the occasion and wanted to speak well. Yet, under the bright lights at burlesque shows, she feels free and able to enjoy performing. On stage she does not need to speak, rather, her body can express her feelings non-verbally.

The cultural phenomenon of burlesque has been the topic of gender and feminist studies (Ferreday 2008; Nally 2009). Contemporary burlesque originated in the mid 1990s in London and New York 'super' nightclubs (Ferreday 2008). Vintage style dress and striptease performances dominate this style as femininities and masculinities are parodied. What sets the contemporary burlesque style apart from early burlesque is the audience, which usually consists of women and gay men (Ferreday 2008). Straight heterosexual men make up the audiences of more traditional burlesque. The nightscape is transformed through the relationship between the performer and audience. This gendered performance, in Cindy's experience, is about a sense of community between the performer and audience. There is a shared set of feminine cultural norms and an awareness that – through classes and performances – particular forms of femininity can be learnt. Creating and transforming queer community spaces through dance has been shown in Detroit for Black and Latina/o LGBTIQ people (Bailey 2014), so too can burlesque.

I have seen Cindy perform several times, including at the launch of a Hamilton Pride festival. She embraces and parodies excess femininity with feather boas, corsets, and sometimes removing her top to show sequinned nipple tassels on her breasts. On other occasions, Cindy completely covers her body in a Pink Panther suit. Burlesque is a vehicle for Cindy to celebrate and destabilise feminine identities. The burlesque subculture is 'associated with a history of women's liberation and sexual freedom that is inseparable from excessively feminine identity performances' (Ferreday 2008, 51). Cindy feels part of a women-centred community, as our following exchange shows:

LYNDA: The nice thing about burlesques is that it's so women centred, isn't it? It's all about women and,

CINDY: Yes, women and the body.

LYNDA: Yes women's bodies.

CINDY: And it's any shape and size. There is no discrimination whatever within the community. All those girls up on that board there [points to images on her wall], they've all very supportive of me. You've only got to look at the way they treat me. They cuddle me and there's a couple there that I really like … It's been a marvellous experience.

LYNDA: Did they have to encourage you to take your top off and wear the tassel, or were you are ready for that?

CINDY: Oh, I was always ready for that. I just wanted to do it. I love the idea. And we had a session there the other night at the beginners' class and, they were teaching the tassel twirling and they all put the tassels on the tops of their bra and I had mine on my nipple. And I took my coat off and did a demonstration. No problem.

LYNDA: Good for you. You are great! That's fantastic.

CINDY: That's a lot of fun. I am telling you. It's a lot of fun.

Feelings of community and belonging is part of what drives Cindy to perform burlesque. An appreciative 'queer' audience and a caring group of performers have opened up more nightscapes for Cindy.

It's also what drives her to be a member of some chartered clubs. The next example comes from a joint interview with Cindy and Sarah, who is transgender, lesbian, in her early 50s, and New Zealand European (see also Johnston and Longhurst 2013; 2016). As discussed in Chapters 3 and 4, both Cindy and Sarah enjoy evenings socialising in clubs, which are semi-private (membership based) bars with restaurants. The clubs attract an older clientele (including returned and retired military service people), who are mostly working class. Sarah and Cindy now feel accepted at their clubs, but not immediately when they both started to live openly as women. Cindy goes to three different clubs.

CINDY: I go to, well I go down to the Cossy [Cosmopolitan] Club, the RSA but my main club is the Workingmen's Club, I'll tell you about that later.

LYNDA: Good, and you talked about the Cubby Hole?

SARAH: I've been going there for four years. They have karaoke there but what I used to like going to was their Ladies' Night, which they used to have every Thursday. Oh, it was wonderful being among all these gorgeous chicks and they didn't know I was a lesbian. It was wonderful! I meant the first night, I was too scared to talk but after I got used to it, it was fine. ... I mean, my voice is obviously male but nobody cared. I would get up and sing anywhere and my voice is male. I've had one person, who is a really good friend of mine actually, who is a member of our [pride] group, and he said to me 'what about your voice?' I don't care, I said.

Sarah spoke with real joy about being out as a woman, and undercover as a lesbian, at the Cubby Hole. She spent many Thursday nights there, singing karaoke and not worrying about her deep voice. In this women-centred nightscape – Ladies' nights on Thursdays – Sarah joined this particular sub-culture of femininity, similar to Cindy's experiences at burlesque performances.

CINDY: I had a friend who is a lesbian, and we had a party one night. She grabbed me and said 'you are coming down [to the Workingmen's Club] all dressed up'. And I said 'OK'. So I went down all dressed up, 'we're going down to the Workies' [laughs] and I went in and oh dear! It was like walking into Antarctica.

LYNDA: Was it frosty?

CINDY: It was very cold in there! But hey, I have been a member for ten years and here I am walking in as a female for the first time and a lot of the guys knew us and of course... oh god! [laughs].

LYNDA: How did they react?

CINDY: Well, they are not allowed to do much in there but you could tell, the eyes say it all... and anyway, I actually got told to tone it down a bit by an ex-cross-dresser that goes in there, and there are a few that go in there that are cross-dressers and one of them I knew told me to tone it down, and I said no, no.

LYNDA: What would that mean, 'toning it down'?

CINDY: Well I was in all white, and I don't think it looked right, but I got dragged in there.

The Workingmen's Club is a conservative space, as Cindy found out when she attended for the first time in women's clothing. She felt her appearance – although not happy with it – was judged more by (ex) cross-dressing patrons than by any other (cisgender) people in the club. This is not surprising, particularly when assessing the way such spaces are cisgendered. Her clothes out her gender identity as transsexual, yet others in the club – particularly (ex) cross-dressers – wish to remain closeted about their embodied identities.

After many return visits, however, Cindy told me that she stopped wearing pants and began wearing dresses and skirts all the time, which she says some people, such as the 'barmaid', now prefer. Cindy explains:

> As Sarah knows darn well, I have made more friends in there now since being female than I ever had before and I have had my old workmates come up to me and slap me on the back and say 'good on ya, I'll shout you a beer' and I have had a couple do that.

Sarah agrees that the atmosphere has changed a lot in the clubs and they enjoy the company of the bar staff but also of the 'ex-railway guys … even the older ones' who appear to recognize that it is not the pants or skirt that matter but 'what's inside'.

Cindy's and Sarah's stories about their clubs and bars relay some difficult situations, but also some positive experiences. This mix of experience illustrates a complex array of negotiations around gendered bodies, spaces and places. Their experiences are also informed by the amount of time that each can spend in either the Workingmen's Club, or Ladies' Night at the Cubby Hole. Both Cindy and Sarah form relationships across genders and this helps establish social bonds, and in turn, troubles cisgender night time spaces (Misgav and Johnston 2014).

Another participant – Sally, who is New Zealand European, female, and aged in her early 70s – shared with me an experience that she had in a 'straight' bar in Melbourne.

> So we were having drink and a meal on South Bank one night and through the glass there were these three guys drinking beer and half way through the meal this guy comes wandering up with a glass of beer in his hands and says 'we want to shout you two ladies a beer if you come and join us'. And I'm sitting there absolutely petrified because the other thing I know about men is that they don't like being fooled. They don't like it. Straight men don't like people like me. They get really aggressive, or can get really aggressive. Luckily the woman beside me, who's a female, said 'oh no we're alright, thank you very much but we're okay'. They kept making faces at us through the window but eventually they just gave up [laughs]. The story is good for my ego but it is potentially dangerous.

Sally feels both elated yet concerned about this experience. She is flattered for the attention by the cisgender men she does not know, yet worried that they may become aggressive. Westbrook and Schilt (2014, 33) in their article 'Doing gender, determining gender' illustrate the cultural ideology that a person's gender is 'authenticated by other people'. This cultural norm naturalises a sex/gender/sexuality system and reinforces heterosexuality as the only desirable (and hence 'natural') sexual form. Sally believes that if the men find out that she is a transwoman, they will accuse her of misleading them. Transwomen are often constructed as dangerous, and this is a discourse used to justify violence against them (Westbrook and Schilt 2009). Some nightscape interactions may produce 'gender panics' where 'people react to disruptions to biology-based gender ideology by frantically reasserting the naturalness of a male–female binary' (Westbrook and Schilt 2009, 34). Within the heterosexualised space of a bar, then, a gender panic may occur if the men challenge what they believe constitutes a woman.

This chapter addressed urban nightscapes to provide insights into constructions of sexed and gendered bodies and spaces. It began by drawing on research about drag queen performances (in Aotearoa New Zealand). In these 'queer' spaces where drag performances occur, gendered bodies intermingle, sometimes producing diverse gender expressions, yet at other times, restricting and normalising gender. Drag is not just about entertainment, rather, it is also about the politics of urban nightscapes. Nightscapes are places where bodies touch other bodies, sometimes creating desire across sex/gender distinctions, sometimes also highlighting the fragility of masculinities and femininities. Queer nightclubs are also spaces where one can find community and feel a sense of belonging. As Vonnie found, they are important sites for takatāpui to find each other, to connect and build support for other community events.

This chapter also addressed 'straight' nightscapes – burlesque shows, Ladies night karaoke, Melbourne bars and private parties. In these spaces transwomen find acceptance, yet they are also places of potential violence and unwanted sexual encounters. A focus on sexed up and gender fluid nightscapes is useful for understanding more about the experiences of gender variant bodies, spaces of inclusion, and power more generally.

8 Transnational and transit(ion)

Documenting bodies, nations and mobilities

The growing field of transgender studies has been challenged to interrogate its whiteness (Haritaworn and Snorton 2013; Roen 2006). Geographers are well placed to add to this critique, and in previous chapters of this book I draw on Aotearoa New Zealand Māori participants' lived experiences. This chapter begins by focusing on some of the lived experiences of fa'afāfine (transwomen) and fa'afatama or 'tomboys' (transmen) in Samoa, before widening the lens to consider the uneven global geographies of human rights for transgender people. The chapter, then, is both national and transnational in scope and it addresses the 'asymmetries of globalisation' (Grewal and Kaplan 2001, 664).

Recent research clearly documents pervasive discrimination against trans people in many countries, including Aotearoa New Zealand, Australia, the U.S. and the U.K. (Bryne 2014). The 2008 New Zealand Human Rights Commission Report 'To Be Who I Am' found discrimination and oppression continues to be felt on a daily basis by those who transgress a binary system of male / female (Liddicoat 2008). The U.K. Equalities Review produced a research project called *Engendered Penalties: Transgender and Transsexual People's Experiences of Inequality and Discrimination* and this is understood to be a significant and comprehensive study on trans people's lives (Whittle et al. 2007). The Australian Human Rights Commission (2009) also conducted a gender and sex diversity project. The *License To Be Yourself* (Bryne 2014) report documents some of the most progressive and rights-based policies and legal rights from around the world. These reports, alongside other research in the U.S. (Amnesty International 2006; Transgender Law Center 2009), provide vital insights into the lives of trans people, their treatment by government agencies and courts. All these projects involved extensive consultation with trans people, and are based on quantitative and qualitative data. They show, at the state level, that trans people are often not provided for in law and policy, and where they are provided for, law and policy is inconsistent and discriminatory.

Moving across borders – for gender diverse people – may throw into question one's state documentation, particularly when migrating to another country (Seuffert 2009). Applying for, and obtaining, consistent documentation regarding gender identity can be a major issue for trans people globally, and raises a number of potential issues for migration, 'including the possibility of suspicion of fraud or withholding of material information' (Seuffert 2009, 449). Not only have key words and concepts – such as mobility, migration, and fluidity – become useful to describe how bodies and places become gendered (Baydar 2012; Cotton 2012; Smith et al. 2016), they are also useful to understand the embodied realities of migration, travel and border crossing. All of these experiences contrast deeply between, for example, sex-working migrants, Indigenous peoples, and HIV+ transgender women.

Gendered citizenship, then, is a form of social identification, which some writers claim is replacing more traditional forms of national identity. Nations often assume their citizens are heterosexual cisgender men or women who conform to 'natural' or 'normal' gender and sex. This chapter troubles this assumption. Some people are excluded from citizenship because of their gender. Citizenship identification – birth certificates, identification cards, and passports – are needed for people's everyday actions, as well as when crossing national borders.

The lived realities of gendered citizenship, and associated feelings of (not) belonging, are highlighted in the first section of the chapter. Here I draw out the social and cultural significance of being excluded from the law for fa'afāfine and fa'afatama in Samoa (Farran 2014). Legal changes affecting the lives of transgender and non-heteronormative people across Pacific nation states are uneven, despite colonial and postcolonial legislative commonalities. By drawing on fa'afāfine and fa'afatama experiences, I illustrate dynamic relationship between culture, tradition, postcoloniality, and legal status. Fa'afāfine and fa'afatama (also known as fa'atamaloa) do not 'fit' recognised legal categories and this has a profound and mundane impact on everyday Samoan life. The second section of the chapter provides an overview of some nations' gender diversity laws and associated human rights where there have been law and policy changes.

Living (with)out legal status

Before turning to an example of the legal rights and experiences of fa'afāfine and, to some extent, fa'atamaloa (or tomboys) in Samoa, I give a brief explanation of some of the gender categories that circulate in and around Samoa. The Samoan words fa'afāfine (plural), and fa'afafine (single) – when literally translated – mean 'in the manner of' or 'like' (fa'a) women (fāfine) or a woman (fafine) (Schmidt 2010). Fa'afāfine identity refers to someone who is assigned male at birth but performs societal duties usually associated with femininity. This is lived out in many ways, including in relation to jobs, having a feminine demeanour, and a preference for the company of women and girls (Besnier 1997; Schmidt 2010; Schoeffel 2014).

Some fa'afāfine are masculine in appearance, while others – particular those living in westernised nations – may use clothing, cosmetics and medical technologies to express femininities, or to occupy a position as neither male or female (Schmidt 2001, 2003, 2010). Regarding the question of sexual orientation (as distinct from gender identity) the 'influx of western concepts of homosexuality has resulted in some fa'afāfine adopting the identity of 'gay man', while simultaneously still identifying as fa'afāfine – although there are also gay Samoan men who are not fa'afāfine' (Schmidt 2017, 3). There are many difficulties involved in describing or even suggesting the socio-cultural spaces and places of fa'afāfine and fa'atamaloa, or tomboys (Tcherkézoff 2014). Another gendered Samoan identity, that has received very little attention, is fa'atamaloa, which is usually understood as referring to:

> girls, or women who are said to be born as girls but who come to be viewed as acting in the way of men at roughly the same stage in life as when 'boys' become *fa'afāfine*. There are two differences between them and *fa'afāfine*. First, they don't claim to be of the other gender: they assert that they are girls, not boys. Second, there is no straightforward Samoan term that designates them as being 'in the way of boys or men'. When Samoans refer to them, they use various circumlocution (eg. 'exhibiting the behaviour of boys or men') or, more pithily, the English borrowing 'tomboy'.
>
> (Tcherkézoff 2014, 115, italics in original)

Tcherkézoff (2014, 116, italics in original) does not attempt to analyse these identities, rather he investigates the 'claims that *fa'afāfine* and *tomboys* make about their own identity and how others talk about them and engage with them'. Tracing the history and use of these identity terms, Tcherkézoff (2014) unravels colonial, Christian and cultural assumptions showing how terms are borrowed and used, as well as highlighting the differences and conflations of genders and sexualities. Recognition and representations of Samoan 'tomboys' – 'a girl or woman who acts like a man in contexts where "strength" (*mālosi*) is particularly central to the definition of manhood' (Tcherkézoff 2014, 116, italics in original) – is rare. Tcherkézoff (2014) notes that the lack of tomboy fa'atamaloa or visibility is due to heteronormative assumptions that tomboys are deemed by many to be not part of custom (aganu'u). Further:

> visitors never see or hear about *tomboys* (there are no shows organised by *tomboys*, for instance), while they will immediately notice the presence of *fa'ateine* [an earlier Samoan term for fa'afāfine] in such highly visible contexts as families, shows, and sports.
>
> (Tcherkézoff 2014, 118, italics in original)

There is evidence, however, that this 'invisibility' is changing. At the 2017 Samoan Fa'afāfine Association (SFA) executive meeting, the president stated:

> This is a very diverse executive council, and I am proud to introduce Vanilla Ice Heather as our first fa'atamaloa member to join SFA. It is our hope for this coming year to embrace all the other culturally diverse members of our society.
>
> (*The Samoa Observer* 2017, no page number)

The very first official gathering (Talanoa) of fa'atamaloa was held in October 2017, and was organised by the SFA. The SFA President applauded the participants for having the courage to attend the Talanoa:

> This gathering is the first of its kind and a historical milestone in the advocacy work of SFA to be inclusive of all gender minorities as stipulated in its mandated functions and Strategic Plan 2016–2020. More importantly is the fact that this is a long overdue process to gauge the perspective and participation of the fa'afatama, a key population under the U.N.D.P. Global Fund Multi-Country Project.
>
> (Luamanu 2017, no page number)

The Talanoa addressed issues such as: building a visible community; the need for safe spaces and support; unemployment; dependency on family for support and the fear of being ostracised from this support system; the challenges of being fa'afatama in a Christian and conservative family environment; and, the difficulty of escaping the expectations and demands of gender roles associated with women (Luamanu 2017). The SFA President So'oalo commented:

> Most associate with other men through sports, friends and bands but rarely with other fa'afatama / transmen. Despite this isolation, culture and norm, the Talanoa session demonstrated their issues and concerns are more similar than diverse. The challenges of growing up fa'afatama are generally suppressed putting first family culture and religion. They are faced with discrimination on a daily basis through verbal and physical abuse generally from close family members and the public … The participants shared stories of surviving abuse and if it weren't for their 'tough' nature and presence, for most suicide was a feasible option. Support varied with some saying it's their mothers or grandmothers who demonstrated the most support, where others confirmed it's the opposite and their fathers were their main supporters.
>
> (Luamanu 2017, no page number)

Terminology was also an issue raised at the Talanoa, with participants encouraging the correct pronouns – he, him, his – for fa'afatama / transmen and it is 'incorrect and disrespectful to call a fa'afatama / transman she, suga or sis. Tomboy is a term that is loosely used and there seems to be a natural and accepted association to the term among their community only' (Luamanu 2017, no page number). Some fa'afatama identify as heterosexual, and some have children.

It is within this social, cultural and political environment that Sue Farran (2014) raises questions about the applicability of western style human rights for fa'afāfine (Samoa) and fakaleiti (Tonga). Farran (2014, 347) notes:

> transgender Pacific Islanders remain beyond or 'outwith' the law in a number of respects and yet also within it when the boundaries shift, especially when the locality in which the law applies and cultural mores that inform the law are taken into account.

She describes some of the geographies of inclusion / exclusions for fa'afāfine according to Samoan domestic laws and notes that any legal changes in countries close to, but beyond, Samoa (plus Tonga) in, for example, Aotearoa New Zealand, may influence Samoan legislation changes.

While Pacific Island nations and peoples are distinct, they also have similarities. When Samoa was colonised it came under English law. Also, in the nineteenth century, Samoa was heavily influenced by Christian missionaries, to the point that Christianity is now completely integrated into the everyday lives of Samoans. Samoa's constitution refers to 'God and Christian principles' with the majority of the current population practising or baptized Christians (Farran 2014, 348). Hence, strong patriarchal and familial structures reproduce many gender inequalities. Yet, due to the Pacific diaspora, many Samoans live in Aotearoa New Zealand, Australia, the U.S. and elsewhere. In 2013, Aotearoa was home for approximately 170,000 Samoans, of which about two-thirds were born in New Zealand (MacPherson 2017). Close ties between countries mean that the 'ebb and flow of people, remittances, news, views, values, and attitudes contrasts sharply with the static nature of the laws that prevail and the very limited legal reform that has taken place since independence' (Farran 2014, 348).

Life in Samoa, therefore, is characterised by both conservativism and the desire for change. For fa'afāfine and fa'afatama living in Samoa, daily life is regulated by customary practices as well as the legal framework, which attributes legal status determined by age, gender, and marital status. There are significant legal consequences when a person does not fit neatly into a western category used in Samoan law – for example wife, husband, child, parent (Farran 2014). Yet in neighbouring postcolonial westernised countries such as Aotearoa New Zealand and Australia, reforms have shifted the way in which the law defines gender, sex and marriage. The meanings and definitions of family, for example, are changing (Melville 2016). The Samoan Prime Minister – Tuilaepa Sailele Malielegaoi – declared that there would be no 'same-sex' marriage in Samoa, despite legislative changes in Aotearoa

New Zealand and Australia (Radio New Zealand 2017). A spokesperson from the Samoa Fa'afafine Association, Tim Baice, said that same-sex marriage legislation is 'not a priority for his group or the Samoan community' (Radio New Zealand 2017, no page number). The Prime Minister – a patron of the Samoan Fa'afāfine Association – said 'as long as Samoa remained a Christian country it would not allow what he referred to as Sodom and Gomorrah practices' (Radio New Zealand 2017, no page number). Baice explains that while some fa'afāfine may be passionate about marriage equality, the group's priorities are:

> The promotion of human rights and the reduction of discrimination and violence based on gender identity, those sorts of projects are their priority projects. They're [fa'afāfine] being very careful about which issues they would like to prioritise so a discussion on marriage equality would just be really out of the ball-park and it could potentially risk all the other sorts of projects.
>
> (Radio New Zealand 2017, no page number)

Some fa'afāfine living in Samoa feel that their status within traditional, colonised social structures grants them considerable rewards and have no great urgency to challenge or change family law (Besnier 2004). Fa'afāfine are members of families and social groups and are hence bound by established cultural codes of conduct. They may be valued members of church choirs, Sunday school teachers, participate in fund raising, and so on (Farran 2014). If fa'afāfine observe 'correct mores, such as church attendance, respect, sobriety, and generosity, which are directed at maintaining [patriarchal] social harmony and equilibrium, then they are generally tolerated' (Farran 2014, 362). Yet, fa'afāfine and fa'afatama may also be rejected by church leaders and family members, and become victims of abuse.

In this legal, social and cultural environment I followed a story that appeared in the *Sunday Samoan* newspaper, the Sunday edition of the *Samoan Observer* newspaper. The newspaper reported on the suicide of a 20-year-old fa'afafine – Jeanine Tuivaikai – and published a full length image of her lifeless body hanging in a Catholic church hall in Apia, the capital of Samoa. The shock and anger over the media item was felt and discussed around the Pacific, and particularly in Aotearoa New Zealand. The news item referred to Jeanine Tuivaika as 'a man' and misgendered her with 'he' and 'his' pronouns. Phylesha Brown-Acton – a prominent leader in Aotearoa's fa'afāfine community – said: 'I am absolutely disgusted by the *Samoa Observer* and their front page photo of a young fa'afafine woman' and the reporting was 'completely inappropriate and disrespectful' (*Pacific Media Watch* 2016, no page number). Brown-Acton went on to say:

> Where is the respect for this young person and her family? The use of such an image to sell newspapers is the lowest form of sales tactics and

the editor and the reporter should be held accountable for such degrading journalism.

(Asian Pacific Report 2016, no page number)

In the face of public hostility the *Samoan Observer* wrote an apology for publishing the image in another controversial item headed up 'And if you're offended by it still, we apologize' (Asian Pacific Report 2016, no page number). It read more like a justification and drew more criticism with Brown-Acton reminding the newspaper – which has the biggest circulation in Samoa – it has 'a track record of misgendering, misclassifying and misrepresenting fa'afafine and continuing to portray and promote fear among the community about fa'afafine' (*Pacific Media Watch* 2016, no page number). The chief editor, Gatoaitele Savea Sano Malifa, then wrote the following 'Apology to our readers' (2016):

> Let me say this is not an easy letter for me to write. Still, I feel duty-bound to write these words, since it is our duty to tell the public we serve, the truth. The truth is that last week, we made a sad mistake when we published a story on the late Jeanine Tuivaiki, on the front page of the *Sunday Samoan*. We now accept that there has been an inexcusable lapse of judgment on our part, and for that we are sincerely regretful. Yesterday, we met with members of Jeannie Tuivaiki's family at their home at Vaiusu, where we extended our sincere apologies, and we are now thankful that we have done so. And so to Jeanine's family we are very sorry. To the LGBT [Lesbian, Gay, Bisexual and Transgender] community in Samoa and abroad, we offer our humble apologies.

(Savea Sano Malifa 2016, no page number)

Hostility towards the newspaper continued (see #BeautifulJeanine) reflecting the backlash towards the historical marginalising legacies of colonial law, Christianity and western interpretations of traditional customs. In the midst of this debate, the President of the Samoa Fa'afāfine Association, So'oalo To'oto'oali'i Roger Stanley, held a press conference and invited all members of the public to join a peaceful parade, starting at the centre of Apia and finishing at the church where Jeanine Tuivaiki was found. Traditional and contemporary discourses entwine in the statement by the Samoa Fa'afāfine Association [SFA]:

> For many years our fa'afafine and fa'afatama community has been silent as we consciously adapt to the norm because Samoa is our home, it is our land, it is our identity and we understand the responsibilities and duties of being Samoan. But these attacks cut us at our very core, and we bleed to the extreme – that, validates our need to break the silence and say Enough is Enough! This is about the environment and the framework we live in – our fa'aSamoa [the ways of Samoa], our aganu'u

[culture], our communities, our Ekalesias [churches], our very own aiga [family beyond parents and siblings] and families, and our own community, the LGBTI community, the laws that govern our behaviour – this is about as massive a wakeup call as any, for all of us to support those members of our communities that are marginalized through existing legislation.

(Samoa Fa'afāfine Association Inc 2016, no page number)

The statement went on to say that religion is used to be both the basis of Samoa society *and* to attack fa'afāfine and fa'afatama. Hence, gender diverse people find themselves in a paradoxical position, never to be fully:

free and equal citizens of Samoa whilst these behaviours and laws are in place. Because these behaviours are legislated, it becomes acceptable to discriminate, and it becomes the norm, which leads down to one path – the victimisation and harm of fa'afafine and fa'afatama.

(Samoa Fa'afāfine Association Inc 2016, no page number)

Laws that inform legal status are shaped by customs, legislation and decisions by judges on individual cases (Farran 2014). In most countries, legal status is attached to the gender assigned to individuals when they are born. Gender is then recorded based on a visual inspection of the newborn's genitalia, and is regarded as a fact that remains significant for the rest of that person's life. There are many laws that operate differently on the grounds of gender and sex:

for purposes of taxation, social benefits, contractual capacity, and property rights. Sex may determine whether a person can commit or be the victim of certain crimes such as rape, infanticide, or sodomy. Similarly, it determines the role that an individual can assume within family law: wife or husband, mother or father, and brother and sister. There are also a number of gender roles in families that do not attract legal consequences but may be shaped by customs or traditional practices, and that in turn may have some bearing on other legal determinations.

(Farran 2014, 355)

Legal rights and obligations, then, based on these morally laden assumptions about gender roles prove to be problematic for gender variant people who do not easily fit a sex / gender category (man, woman, girl, boy, etc). When countries reform these laws and adopt, for example, gender neutral language and make specific provision for gender variance, then some of these legal dilemmas are resolved (Farran 2014). Samoan law does not prohibit discrimination on the basis of gender identification and sexual orientation, yet the 1960 Samoan Constitution states that: 'All persons are equal before the law and entitled to equal protection under the law' (Article 15[1]), prohibits discrimination on the grounds of 'descent, sex, language, religion, political or other opinion, social origin, place of birth, family status, or any of them' (Article 15[2])' (Farran 2014, 356). Gender and sex are implicated within intersecting structural inequalities because of one's ethnicity, social class, rank, and so on. The tragedy of the young fa'afāfine woman Jeanine Tuivaikai who took her own life in the catholic church hall could be understood as symbolic of not belonging to a nation state or state sanctioned institutions, such as the catholic church. The next section considers nations, and moving between nations, with respect to citizen rights.

Documenting sex and gender within and across nation states

'Transgender is like a refugee without citizenship' (Bird 2002, 366). Trans activists, supporting organisations, and scholars are mapping the connections between gender variant people's rights and nationalism. Scholarly work has emerged in Hong Kong (Emerton 2006), the U.K. (Aizura 2006; Munro and Warren 2004), Australasia and the Pacific Islands (Alexeyeff 2000; Besnier 2002, 2004; Farran 2014; Mageo 1992; Schmidt 2001, 2003, 2010), and the U.S. (Currah and Minter 2000; Kirkland 2006; Spade 2011), for example. The concern of these studies is the reality that when people are unable to align their gender with official state documentation they are excluded from many aspects of citizenship.

Gender diversity within nation states needs to be recognised and respected. The right to self-determination of transgender people – as enshrined in the Yogyakarta Principles (International Commission of Jurists (ICJ) 2007) – means that there is a legal pathway for lawmakers to follow that allows self-declarations about gender identity without any discriminatory preconditions. The right to self-determination helps gender variant people achieve positive physical, social, mental well-being and health. If one is asked to repeatedly produce identity documents that do not align with name and appearance then the risk of experiencing abuse, discrimination, prejudice, and harassment intensifies (Szydlowski 2016).

The list of countries deemed to have 'trans-friendly' laws is growing slowly yet still the vast majority of gender variant people cannot obtain official documents that match their name and sex with gender identification. In countries where people do have access to legal gender recognition, individuals can still face significant challenges. Many countries violate fundamental human rights when they do not recognise gender variant people or make compulsory medical diagnoses or interventions, including sterilisation, in order to receive official identification documents.

Argentina made world news by becoming the first country in the world to legalise the right to self-determine one's gender. The Gender Identity and Health Comprehensive Care for Transgender People Act of 2012 makes it possible for people to request that their recorded sex, image and first name are changed to match their gender identity. Furthermore, no medical opinion or evidence is required (Bryne 2014). The law states:

> Article 1 –*Right to gender identity.* All persons have the right:
>
> 1   a) To the recognition of their gender identity;
> 2   b) To the free development of their person according to their gender identity;
> 3   c) To be treated according to their gender identity and, particularly, to be identified in that way in the documents proving their identity in terms of the first name/s, image and sex recorded there.
>
> (http://tgeu.org/argentina-gender-identity-law/)

The definition of gender identity is:

> understood as the internal and individual way in which gender is perceived by persons, that can correspond or not to the gender assigned at birth, including the personal experience of the body. This can involve modifying bodily appearance or functions through pharmacological, surgical or other means, provided it is freely chosen. It also includes other expressions of gender such as dress, ways of speaking and gestures.
>
> (http://tgeu.org/argentina-gender-identity-law/)

The Gender Identity Act goes on to explain that minors may request to change their recorded gender. However, these requests must be made through their legal representatives (http://tgeu.org/argentina-gender-identity-law/).

In a study involving 452 Argentinean transwomen, it was found that 260 (57.5 per cent) had, within the first 18 months of the law, obtained a new gender identity card (Socías et al. 2014). The study shows that the most 'empowered' transwomen were among the first to take advantage of the law.

Malta has arguably the most inclusive legal framework due to the passing of the historic Gender Identity, Gender Expression, and Sex Characteristics Bill (2015) adopted by their parliament on 1 April 2015 (see Malta Gender Identity, Gender Expression and Sex Characteristics Act (2015) and the Bill here: http://justiceservices.gov.mt/DownloadDocument.aspx?-app=lom&itemid=12312&l=1). Malta's policy allows their citizens to change their gender identity on official documents – passports, birth certificates – without medical approval or treatment. Maltese citizens may choose an X or 'decline-to-state' option for passports. Malta's laws go further by actually prohibiting requests for medical information about gender change applications. The Bill regulates healthcare provision and prohibits 'normalising' genital surgeries on intersex infants. The Transgender Europe Co-Chair and lawyer, Alecs Recher (2015, no page number) applauds the Act:

> The Gender Identity, Gender Expression and Sex Characteristics Act sets a new benchmark for Europe. We are thrilled about the respectful, comprehensive and yet practical aspirations of this new Maltese act. It is firmly built on trans and intersex persons' right to be recognized for who they are. Furthermore, the GIGESC Act creates the conditions for an equal society as it recognises and protects trans and intersex persons in all spheres of life.

Also in Europe, Denmark's laws allow for self-determination following the passing of the Danish Gender Recognition Law in 2014. The new law abolishes requirements for any medical interventions (such as psychiatric diagnosis, hormonal treatment and sterilisation) and has in place an easy-to-navigate administration procedure for people to 'receive a new gendered social security number and matching personal documents such as a passport, driving license and birth certificate in accordance with the new gender' (TGEU 2014, no page number). There is, however, a waiting period of six months for applicants, and a minimum age requirement of 18.

> Under the new law, applicants for gender recognition have to refresh their application after a waiting period of six months. According to Danish lawmakers, this measure was introduced to prevent persons from making hasty decisions they would later regret. However, this **imposed delay in the procedure prevents trans people from changing their documents quickly when necessary**, for example when applying for a job, travelling internationally or enrolling in education. Furthermore, TGEU is concerned that the waiting period may also **perpetuate misconceptions of trans people as being "confused"** about their gender, instead of encouraging them to change their documents quickly so that they can participate fully and freely in all aspects of life.
>
> (TGEU 2014, no page number, bold text in original)

These shortcomings can also be seen in other European countries. Ireland has adopted a progressive gender recognition law. In 2015 the Irish Senate ended a long wait for transgender people's recognition. People over the age of 18 years are allowed to 'self-declare' their gender without any medical criteria as part of the application. The law, however, is restrictive for 16 and 17-year-olds who do require certification from medical practitioners, parental consent and a court order (TGEU 2015). There is no provision in the law for people under 16 and anyone with a non-binary gender identity, plus intersex people, do not benefit from this law (TGEU 2015). Transgender Equality Network Ireland (TENI) campaigned for many years for the introduction of the legislation, noting that the 'struggle started when Dr. Lydia

Foy applied in 1993 for a change of her birth certificate' (TGEU 2015, no page number).

The United States National Centre for Transgender Equality (NCTE), founded in 2003, has a focus on advocacy and education issues at the national level (Bryne 2014). Some of their work has helped bring about the elimination of proof of 'sex reassignment surgery' in order to update one's passport details. People whose birth certificate does not show their preferred gender can get a form that states they have had 'appropriate clinical treatment for gender transition' (Bryne 2014). This vague wording means that the U.S. State Department is not privy to any treatment as the documentation discourages medical professionals from providing in-depth medical information, which goes some way towards protecting the applicant's privacy. The policy also instructs staff to be respectful, ask only relevant questions, and to use the person's preferred gender pronouns (U.S. State Department Foreign Affairs Manual 2016). The NCTE use a number of strategies to make progress with U.S. policy. In particular, they frame these advances around scientific and medical issues, and avoid making claims under human rights legislation. In this way, the U.S. Department of State instigated the policy changes so that they aligned with the World Professional Association of Transgender Health (WPATH, see www.wpath.org).

Also in the U.S., trans people in couple relationships do not have to become single or divorced in order to obtain legal gender recognition. Surprisingly, even though many states do not recognise same-sex marriages, when one spouse in a different-sex marriage undergoes gender transition their marriage remains valid (Bryne 2014).

Turning to countries in the Southern Hemisphere – particularly Aotearoa New Zealand – passport details, but not birth certificates, can be changed without needing medical information or a medical diagnosis. The threshold, then, for changing one's gender and sex details, is higher for birth certificates than it is for passports. The Human Rights Commissions for both Aotearoa New Zealand and Australia recommend that the process be simplified. Applications to change birth certificates are made to the Family Court with separate provision for people under the age of 18 in Aotearoa New Zealand (unless aged 16 or 17, they have been married, or are in a civil union or de facto relationship at which point one is deemed to be an adult). Legislation focuses on evidence of medical interventions (for adults) yet for minors, and if the Family Court agrees, evidence may be in the form of medical treatment the minor will have to change sex (Bryne 2014).

As a result of the New Zealand Marriage (Definition of Marriage) Amendment Act 2013, couples are not required to divorce or dissolve a civil union when official documentation is changed. The Act now defines marriage as 'the union of two people, regardless of their sex, sexual orientation, or gender identity' (Marriage (Definition of Marriage) Amendment Act 2013).

There are some countries that legally recognise a third gender / sex option, for example, and as noted above, it is possible to register a third gender in New Zealand by having 'indeterminate / unspecified' on one's passport once medical transitioning has started. It is also possible to complete a statutory declaration and request X on the passport (and prior to the X option, passports featured a dash (–) (International Civil Aviation Organisation Guidelines 2014). Bryne (2014) notes that Nepalese adults are issued Citizen Cards. Up until 2007 the card was frequently denied to those who wished to register as a third gender, i.e. not male or female (also known as 'meties' in Nepal). While the courts have established the right to use a 'third gender' category, the implementation process has been slow. In Nepal, a medical diagnosis is not always required for meties in order to receive an 'Other' Citizenship Card. Similarly, hijra in India do not require a medical diagnosis to change sex / gender documentation, yet the legal status of these documents remains unclear (Bryne 2014).

In summing up gender identity recognition laws, Bryne (2014, 9) notes:
Ideally, progressive laws and policies will:

- Be based on self-defined gender identity rather than verification by others;
- Include more than two sex / gender options for those who identify outside the binary categories of male and female;
- Include intersex people;
- Apply to all residents, including those born overseas;
- Link to broader human rights, particularly access to health services that enable someone to medically transition if that is their choice;

And will not:

- Require a medical diagnosis of gender identity disorder, gender dysphoria, or transsexualism;
- Require transition-related medical treatment, such as hormonal therapy or gender affirming surgeries;
- Require sterilization, either explicitly or by requiring medical procedures that result in sterilization;
- Require living continuously or permanently in one's gender identity;
- Require divorce or dissolution of a civil partnership;
- Prohibit parenting now or the intention to have children in the future;
- Be governed by age restrictions. Options for children and youth should recognize their evolving capacities.

In short, the process for changing one's gender / sex and name should be simple, done quickly, and low cost (preferably free), transparent and completely confidential.

What the above cursory glance at different countries' gender identity laws, and participants' border crossing experiences, tells us is that across the globe legal recognition and human rights are patchy for transgender, gender variant and intersex people. Each nation state has different laws regarding who and what counts towards being able to claim citizenship. The importance of establishing civil identities motivates activists to fight for legal recognition of gender and sex, rights to marry (or stay married) regardless of gender, and the right to be an indeterminate gender or intersex.

I asked all of the gender variant geographies' research participants what they thought of their legal status in Aotearoa New Zealand and in the United States. The participants in Aotearoa New Zealand were, on the whole, happy with their legal rights and legal status. It should be noted, however, that the lack of gender identity data from countries' censuses prevents both the estimation of transgender population numbers, and hence support and health services that may be needed. The Aotearoa New Zealand Government developed a policy tool aimed to provide a statistical standard for gender identity data collection (Pega et al. 2017). The *New Zealand's Statistical Standard for Gender Identity* (New Zealand Statistics 2015, 4) includes the following classification of gender identity:

1   Male

   11   Male/Tāne

2   Female

   21   Female/Wahine

3   Gender diverse

   30   Gender diverse not further defined / Ira tāngata kōwhiri kore
   31   Transgender male to female / Whakawahine
   32   Transgender female to male / Tangata ira tāne
   39   Gender diverse not elsewhere classified / Ira tāngata kōwhiri kore.

This is the first of its kind, globally, and the implications stretch to every country and most organisations. Pega et al. (2017, 220) claim that incorporating the standard into 'health records, population surveys, and the census will improve national population prevalence and other demographic estimates and health services and outcomes monitoring'. With this statistical evidence researchers will be able to design, cost, implement and evaluate health policies and services for transgender populations. The policy is not perfect, however, as it fails to include a question based on sex assigned at birth and current gender identity. There are many transgender people who identify within the gender binary as male or female, and not as gender diverse. Further, the policy has no provision for people who are intersex. Yet, despite these limitations the policy is hailed as an innovative way forward.

The census – how it is constructed, who is included / excluded – is closely related to legal changes pertaining to relationship status. Amelia, who is aged

18–24, NZ European, MtF, and pansexual, talked about how she felt Aotearoa New Zealand's Marriage Amendment Bill:

AMELIA: I love the fact that you called it the 'marriage amendment bill'. So many people just call it the gay marriage bill, right?

[We both laughed]

AMELIA: It annoys me every time I hear someone say it. Like it was even on my university lecturer's page. I was like, I literally crossed it out on my page where they wrote it. It's not [a gay marriage bill]. It's more than that. It's the ability for a trans person to be married to their partner.

Amelia went on to say that changing one's birth certificate in Aotearoa New Zealand was difficult.

> Changing your birth certificate is actually really difficult. You need to have had surgery or been on hormones for a very long time then you go to court. You need to send in all the paperwork and if that, if that fails you need to go to court. And if that fails you're screwed. You can't again, I don't think, or for at least not for about five years.

In another research project involving group members of the Hamilton Pride society, Cindy and Sarah explained to me that it is was possible to be designated X instead of male or female in New Zealand passports. Cindy said that X 'means your sex is undetermined'. One is neither female or male, transwoman nor transman. While this move to using X is useful for people who may be intersex, or genderqueer, Cindy felt it was not always easy to be 'X': 'it's not ideal because a lot of computers will not recognize it.' Sarah adds: 'It's not totally international.' Cindy went onto explain an experience she had when returning from Sydney, Australia to Auckland New Zealand: 'The computer wouldn't compute the X and it wouldn't issue a boarding pass.' A border control officer was called to rectify the issue and Cindy has since changed her passport to F for female (see Doan, 2010, 637, on passports; and Johnston and Longhurst 2016).

It is sometimes these in-between states – neither transman or transwoman – that perplexes not only lawmakers but some transgender people too. One of the participants, Emily, who identifies as female, is aged in her early 40s, and Asian ethnicity, was particularly annoyed with some naming decisions. Emily is concerned with people who:

> don't change their name and who go by the same name, [and] who claim to be a woman, and people who don't get [medical] treatment and then they … use one letter, one letter, as a first name, made out of their given name, but they don't get their IDs changed.

I asked her to elaborate:

LYNDA: So they won't change their ID, they won't change their passport, they won't change anything?

EMILY: No. But of course they also look queer so they are different. But this friend of mine who does that, doesn't change their passport and ID, insists that we use the female pronoun and just this name that she made up. I never said anything because it's politically incorrect to say anything but deep down I just felt like, at every condition like, she seemed to mock me and everything I have to go through and I don't have the same choices as she does because if she's harassed she easily go back to male because she is more queer and sort of like Brad Pitt with long hair, longer hair, yes but could be like him in an instant. So she kind of traverses between male and female.

LYNDA: Like she can decide where to go with that?

EMILY: Right but I don't have that choice, like not my choice and right and so.

LYNDA: It's a different position to be in?

EMILY: We feel threat differently, as in airport security line and stuff. Like um, it's okay for somebody like her, her looks ... the ID and the queer presentation or stares, but that's it. Because she can always, yeah, she can go back, go back, but it's different for me. So that's why all the [physical] changes were so important for me. The mismatch would be devastating and actually quite dangerous.

It's worth pausing here to consider further the implications of gender variance and embodied experiences in light of state-sanctioned classification and management of people based on gender (Herman 2015). Emily has changed all of her legal documentation, and has had many surgeries, in order to live freely as a woman and without question from authorities. She refers to a friend who is in-between genders and resents the fact that people with fluid gender identities can be mobile, both with their appearance and their movements around the globe. Emily's own gender stability is not guaranteed when her friend continues to be gender fluid. Indeed, it's worth remembering that transgender, as a category, was originally intended to be an inclusive term and the basis for political activism. Yet, the category 'transgender' now has ontological functions due to the way in which it has been adopted by social services (see Davidson 2007, and Valentine 2007).

Emily talked, at length, about airports. In a post 9/11 era state security and surveillance is intense, particularly at U.S. airports. Any perceived identity inconsistencies may be viewed as a national threat. Bodies are 'red-flagged' (and deemed 'dangerous') at airports in light of any identity inconsistencies. For transgender people 'travel by way of airports demands meticulous negotiation of disclosure and non-disclosure in order to avoid being singled out as a security threat for inhabiting an abnormally gendered and inconsistently documented body' (Herman 2015, 87). In the U.S., the National Center for Transgender Equality (2014) recommends that transgender people carry letters from medical staff that attest to any transgender-related medical issues, or that travellers complete a notification card (usually used to disclose medical conditions or disabilities) in order to diffuse any identity problems (Herman 2015). The *Know your rights* document (National Center for Transgender Equality 2014, 1–2) states:

> The name, gender, and date of birth must match the government-issued photo ID you will provide when passing through security. The Secure Flight program checks this information against government watch lists, and gender information is used to eliminate false matches with the same or similar names – not to evaluate a person's gender. If you have different names or genders listed on different ID, you can choose which to provide, so long as you bring photo ID that matches your reservation. TSA Travel Document Checkers will check as you enter security to ensure that information on your ID matches your boarding pass. It does not matter whether your current gender presentation matches the gender marker on your ID or your presentation in your ID photo, and TSA officers should not comment on this.

Emily is aware of these requirements, so I asked her if she still feared security checks at airports. She replies:

> I have a lot of fear especially after I got my new ID and passport and travelling for the first time. I didn't know what would happen even though my doctors had certified it. It's so strange that it has to be born certified [laughs]. Most cismen and ciswomen don't have to be born certified, even though they have deviations. So that's my feeling … like a disabled body, having to move through a space like that. But like I get into Kentucky, of all places. I was very nervous but nothing happened so after a few times I felt like my confidence got boasted and I don't think about it as much. But in the beginning I felt self conscious as if I was an imposter. That was my feeling, yes, with correct IDs and everything. But I had, I carried this letter from my psychologist, it's called a carry letter, basically states my medical history about my transition so in case anyone … I've never had to use it, but just having to carry that is something like, it's like a torture.

Emily reflects on the way she negotiates airports and extra documentation. Ideally, she does not want to be in a position where she has to show her letter to prove that she is a woman. If she is forced to show her letter, she will be outing herself as 'abnormally bodied, rather than attempting to "pass" through security unnoticed' (Herman 2015, 87). At this point, transgender bodies become vulnerable to detainment and identity profiling. Emily is not white, hence her ethnicity may also increase suspicion and closer scrutiny than others. Emily told me that she carried her letter in her purse, at all times, making her consider what it must have been like for Jewish people who had to carry identification during the Holocaust. She admits that over time she forgot to carry the letter, and the 'oh my god, I don't have the letter' panics subsided. Emily says: 'I begin to grow into this body, just now, so that's the process.'

On an international journey, Emily's identity as a woman was confirmed when she was mistaken for being a young female student, while one of her actual pupils – 'he's white, he's older, he's bald' – was deemed to be the tutor. Here is an example of patriarchal and racist assumptions that deem bald, older, white men to be authority figures while young-looking Asian women 'must be' students. During this incident, Emily was relieved that there was no (trans)gender trouble, yet she was also 'worried that they would pull up the old data or whatever it is, but there was no trouble at all'. A border security officer did take her passport away and questioned the need for a student study visa in the country she was visiting. All of these intersecting 'identity inconsistencies' were created by border security officers. Transgender border issues are intersectional and not confined to gender alone.

In this chapter I have addressed the ways in which gender variant bodies are documented, legalised, and (not) counted as citizens. One's legal status is intimately connected to one's everyday lived experiences and feelings of (not) belonging to place. For fa'afāfine and fa'afatama in the Pacific Island state of Samoa, the lack of legal status takes on social and cultural significance. What it means, then, to document bodies, nations and mobilities is highly variable from one place to the next, often contradictory, and is porous and unstable. Research on transgender people's legal and citizen status demands a multi-scalar approach in order to understand the lived and embodied experiences in complex global and local dynamics.

Countries that commission and publish reports on the status of transgender people, show that trans people are often absent from laws and policies. Furthermore, in examining particular countries' policies and laws that include gender variance, one can see a great deal of inconsistent and discrimination. Mobility may be restricted and crossing borders can cause panic for some gender diverse people.

Activists across the globe continue to advocate for progressive laws and regulations on gender recognition. Legal documentation – such as birth certificates, passports, and government-issued identification (ID) cards, drivers licences or voter registration cards – are all bound up in the jurisdiction of nation states. These legal state documents are required for day-to-day living, such as: accessing healthcare; employment; social welfare; opening a bank account; and, enrolling in educational institutions. Successful legal change will happen when nation states adopt policies and laws that include gender variance. This local–global nexus informs people's daily lives and their feelings of (not) belonging.

9 Conclusion

Beyond and back to binaries

As I write this conclusion, the Aotearoa New Zealand Government statistician declared that questions on New Zealanders' sexuality and gender diversity will **not** be included in the next census because statisticians could not 'make them work'. As noted in the previous chapter, the New Zealand Government was breaking new ground, and certainly subverting the dominance of the male / female binary, by including a set of questions about gender diversity in the country's five-yearly census. The census was, finally, moving away from the sole mandatory male / female question by allowing New Zealanders to choose one of six gender identity options. Aotearoa was leading the world, with the promise to create positive change around the globe! It was to our surprise that the New Zealand Government reversed their decision to incorporate this policy in the 2018 census: 'Statistics New Zealand ran a series of tests on a third gender option and found that erroneous or deliberately inaccurate answers made the data unreliable' (Cooke 2018, no page number). Further, the Government Statistician said they were 'disappointed that after public trials, they could not find questions that would lead to accurate, consistent, and useable data' (McDonald 2017, no page number).

Much outcry was heard from various rainbow, queer, and trans individuals and groups. Toni Duder, from Rainbow Youth, a charity that supports gender and sexually diverse young people, says:

> without the basic statistics to inform numbers around our community, support and education around the rainbow community will continue to be underfunded and under prioritised. This then leads to the negative health outcomes we see in the rainbow community. It's not good enough.
>
> (McDonald 2017, no page number)

Aych McArdle, another queer activist, joined the public outcry: 'if you don't count someone, you're almost saying they don't count', and questioned: 'is it, at worst, institutionalised homo/trans/intersex phobia?' (*Newshub* 2018, no page number). Obviously, transgender people are not a fixed or homogenous group, and as this book shows, there are multiple gender and sex expressions and identities. This multiplicity was too much for Aotearoa New Zealand Statistics, who, in the short term, will not queer the census (Doan 2016).

Needless to say, this example highlights how the nation moved beyond and back to binaries. The cisgender male / female binary is articulated, disrupted and implemented through bodies, places and spaces. For transgender people, the decision of Statistics New Zealand to not include gender diversity census questions triggers feelings of outrage, precarity and estrangement. Transgender people's emotionally charged experiences of discriminatory practices are spatial, that is, felt at the level of the body, and as this example shows, at the scale of the nation. The promise, then withdrawal of, a census that gathers gender diverse information is indicative of the way power circulates in and through bodies and places.

The aim of this book was to illustrate that in order to understand the production and expression of gender variant people's lived experiences – whether they be female transsexual; transgender woman; female; female MtF; female transsexual; woman MtF; transfemale; transmasculine; genderqueer; intersex; intersex trans male, tatatāpui, non-binary, fa'afāfine, fa'afatama, tomboys, agender, or anything else – one must also pay attention to space and place. Throughout the book I filter gender variant bodies through a geographical lens. Participants' life experiences in relation to places and spaces are presented as multi-scalar in order to prompt new ways of thinking about co-construction of gender, sex, place and space.

Transforming Gender, Sex, and Place asserts that despite decades of troubling gender, geographers are only just beginning to consider trans and gender variant subjectivities (Browne et al. 2010). It needs to be acknowledged that the field of transgender and gender variant studies is, itself, geographical and flourishing in North America and Europe (Aizura et al. 2014). Research is beginning to appear that goes beyond Anglophone western countries. Scholars are showing the way in which transgender knowledges and practices move across borders, regions, rural–urban spaces and nations. Yet, it must be acknowledged that often highly medicalised categorisations such as 'transsexual' are very mobile, while culturally specific terms such as 'fa'afāfine', travel less distance. It also needs to be acknowledged that notions of what being a transgender person *is* may conjure up other concepts such as metropolitanism, eurocentrism, whiteness, cosmopolitanism, globalisation and transnationalism. A 'queer methodology must facilitate telling and interpreting narratives that do not inadvertently impose meanings rather than seeking to rework and create new narratives' (Gorman-Murray, Johnston and Waitt 2010, 101). The book incorporates queer methods in order to remain open to difference, and to always destabilise and question practices that have come to be valued for knowledge production. Transgender knowledges matter, as well as increased representation of transgender voices in publications and as publishers (Veale 2017).

A number of provocative, political, and activist lines of research are opening up on transgender, intersex, and gender variant geographies. Much of this research is informed by trans, queer and feminist understandings of gendered embodiment, places and spaces. There is room to elevate the importance and distinctiveness of gender variant people's geographies within LGBTIQ+ research. Geographers are yet to consider, in any depth, the normative and privileged places associated with being a cisgender person. By examining the co-production of gender, sex, place and space we can enrich our understanding of the ways in which gender variant people feel in, and / or, out of place. Participants' stories, media items, my own experiences, and research illustrates the mutual constructions of transgender bodies with homes, bathrooms, activist and community spaces, workplaces, urban nightscapes, and nations. These are powerful and uneven gender variant geographies. Indeed, scholarship on theoretical discussions, representations, and social constructions of bodies are enriched by the 'real' fleshy lived and visible (Namaste 2000) materialities of bodies. By offering an explicit reading of transgender and gender variant people's bodies, spaces and places, I highlight often overlooked exclusionary spaces and cisgender privilege.

This book considers a number of spaces and places. Transgender, queer, feminist, and poststructuralist theories are used to argue that transgender bodies and places are discursively and materially produced. Chapter 1 sets out this theoretical 'toolkit', which is utilised throughout the book. I argue that one cannot separate representations from lived realities, theories from actions, and bodies from places. These binaries are mutually constitutive. The chapter canvases research that seeks to deconstruct the binary framings of man / woman and male / female roles, spaces and places. Furthermore, there is evidence that the queering of genders (and the relationship between gender and sex) has meant that geographers and other scholars are now tracing the ways in which bodies experience and live their gender beyond normative binaries.

For a number of years transgender scholars have been calling for a focus on everyday lived and embodied experiences (Prosser 1998). Chapter 2 focuses on the body. Bodies are the centre of personal identity formation and hence the fleshy materiality of bodies, including all gender and sex characteristics, are central to self-identification. This chapter highlights people's feelings and memories of the places where they gain a sense of gender and sex identity. Close attention is paid to language used as this helps illustrate the many ways gender variant people feel both in and / or out of place. As noted elsewhere, one's gender and sex cannot be separated from sexuality, class, race, ethnicity, age and dis/ability.

Home takes on many forms, and feelings of home can be found across spatial scales. Chapter 3 examines home as a private, family, or whānau place. Transgender, intersex and gender variant people's senses of 'home' operate differently across diverse populations. The very design of domestic home space may affirm and disrupt cisgender and heteronormative bodies and relationships. A sense of home is often linked to one's gender coming out story, and inevitably, to important familial relationships.

The micro spaces of toilets, bathrooms, restrooms and lavatories are considered in Chapter 4. Binary gender bathrooms remain a persistent problem and continue to reinforce a limited understanding of embodied biological differences and gender expressions. Unnecessarily gendered public bathrooms – as single rooms – can be subverted. The chapter shows that when public institutions include 'all gender' bathrooms, some cisgender people reach and 're-gender' these toilets. Within the private domestic spaces of homes, bathrooms become important places of bodily maintenance, self-surveillance, and refuge, yet also sometimes assault.

Thinking about the spatial formation of activist communities furthers understanding of the politics of resistance. Chapter 5 addresses transgender activism through an embodied and intersectional lens. Experiences and politics of LGBTIQ and takatāpui activism differs, as can be seen at gay pride events. The relationships between genders, ethnicities, sexualities, and classes need to be acknowledged, rather than hidden, under the rainbow banner at gay pride events. Chapter 5 shows that transgender activism occurs beyond queer spaces, and in a myriad of ways, in rural places. There is potential for gender variant activism to radicalise conservative places and hegemonic institutional spaces.

Workplace experiences, as discussed in Chapter 6, illustrate the ways in which binary gendered spaces matter, particularly when transgender people have work experiences both pre and post gender transition. Gender variant people's workplace experiences show how income precarity – as it is embodied, felt, and encountered – is managed. A great deal of self-care, collegial support, and the ability to construct a workplace where one does belong, is vital. Chapter 6 shows that workplaces may be both challenging and inclusive places for gender variant employees. What is clear – in nearly all of the spaces canvased – is that there is an absence of organisational policies and practices that account for gender variant people's experiences. Rather, work organisations continue to imagine bodies and spaces as normatively gendered.

City nightscapes provide spaces where gender diverse bodies intermingle in intimate ways, as shown in Chapter 7. Here many transgender people may feel welcomed, yet, there are also some nightclub spaces where gender expressions are restrictive and normalising. In short, there are gender politics in city nightscapes, with some transgender people feeling more at ease in 'straight' bars, than queer bars. Queer bars may be, however, important community spaces where – in Aotearoa – takatāpui connect with each other and build supportive alliances. These intersectional spaces – the mutual construction of genders and ethnicities – are found in urban entertainment nightscapes, both public and private.

The social, cultural and legal gender variant geographies of Pacific people are uneven, despite (post)colonial commonalities. In Chapter 8 I illustrate the dynamic relationship between cultures, traditions, and fa'afāfine and fa'afatama or tomboy experiences in Samoa. When bodies do not 'fit' state sanctioned legal categories, there are implications for daily life in Samoa, and beyond. A few nations grabble with how to represent transgender bodies in census questionnaires and formal legal documentation, such as birth certificates, passports, and government issued ID cards, driver's licence or voter registration cards. Many nations do not, however, seek to change legal documentation and rather choose to stay with a cisgender normative male / female binary. For many gender diverse people, being absent from a nation's documentation means feeling excluded, or, 'not worth counting'. The everyday lived experience of being 'absent' is thrown into stark relief when people are misgendered, misclassified and misrepresented. Laws and legal documentation do not stand alone; they are also shaped by social and cultural customs.

A new kind of gender variant politics is emerging, one that is committed to trans, feminist, and queer theories, as well as place and space. The book includes people's experiences of living beyond metropolitan areas, away from cosmopolitan spaces. Ultimately, I hope this book shows some of the diversity and complexity of embodied experiences of gender variance and that transgender bodies are constructed in and through different places. There is a promising and hopeful future for gender variant geographical investigations. This book, then, is just one small step towards showing dynamic gender variant geographies, where all spatial boundaries are porous and permeable.

Transforming Gender, Sex, and Place has attempted to show that gender variant geographies are filtered through people's embodied feelings of belonging and alienation. Claiming and making inclusive gender variant places, is an ongoing and transformative project and one continues to revolve around, resist, and live in-between and beyond binary gender.

References

ABC News (2014). *Gender ruling: High Court recognises third category of sex.* 3 April 2014. [Online]. Available at: www.abc.net.au/news/2014-04-02/high-court-recognises-gender-neutral/5361362 [Accessed 4 April 2014].

Ahmed, S. (2004). *The cultural politics of emotion.* New York: Routledge.

Aiello, G., Bakshi, S., Bilge, S. Hall, L. K., Johnston, L. and Kimberee, P. (2013). Here, and not yet here: A dialogue at the intersection of queer, trans, and culture. *Journal of International and Intercultural Communication*, 6(2), pp. 96–117.

Ainsworth, T. and Spiegel, J. (2010). Quality of life of individuals with and without facial feminization surgery or gender reassignment surgery. *Quality of Life Research*, 19(7), pp. 1019–1024.

Air New Zealand. (2008). Pink Flight promo video [Online]. Available at: www.youtube.com/watch?v=bfhZ76SyVc0 [Accessed 29 June 2009].

Aizura, A. (2006). Of borders and homes: The imaginary community of (trans)sexual citizenship. *Inter-Asian Cultural Studies*, 7(2), pp. 289–309.

Aizura, A. Z., Cotton, T., LaGatta, C. B/C., Ochoa, M. and Vidal-Ortiz, S. (2014). Introduction. *TSQ: Transgender Studies Quarterly*, 1(3), pp. 308–319.

Alexeyeff, K. (2000). Dragging drag: The performance of gender and sexuality in the Cook Islands—dancing. *Australian Journal of Anthropology*, 11(3), pp. 297–308.

Amnesty International (2006). *Stonewalled — Still demanding respect!: Police abuses against lesbian, gay, bisexual and transgender people in the USA.* [Online]. Alden Press. Available at: www.amnesty.org [Accessed 24 October 2017].

Andersson, J., Vanderbeck, R. M., Valentine, G., Ward, K. and Sadgrove, J. (2011). New York encounters: Religion, sexuality and the city. *Environment and Planning A*, 43, pp. 618–633.

Andrucki, M. J. and Elder, G. (2007). Locating the state in queer space: GLBT non-profit organizations in Vermont, USA. *Social and Cultural Geography*, 8(1), pp. 89–104.

Argentina Gender Identity Law (2012). *English translation of Argentina's gender identity law as approved by the Senate of Argentina on May 8, 2012.* [Online]. Available at: http://tgeu.org/argentina-gender-identity-law/ [Accessed 1 March 2017].

Asian Pacific Report (2016). *Sunday Samoan* condemned for 'disgusting, degrading' reporting of death. [Online]. Available at: https://eveningreport.nz/2016/06/20/sunday-samoa-condemned-for-disgusting-degrading-reporting-of-death/ [Accessed 22 June 2016].

Austin, M. (2016). Burlesque: Glamour, decadence and style. *Stuff.co.nz*, 5 November. [Online]. Available at: www.stuff.co.nz/entertainment/stage-and-theatre/85975211/burlesque--glamour-decadence-and-style [Accessed 7 November 2017].

Australian Human Rights Commission (2009). *Sex files: The legal recognition of sex in documents and government records.* Human Rights and Equal Opportunity Commission. [Online] Available at: www.humanrights.gov.au/our-work/sexual-orientation-sex-gender-identity/publications/sex-files-legal-recognition-sex [Accessed 13 November 2010].

Bailey, M. M. (2014). Engendering space: Ballroom culture and the spatial practice of possibility in Detroit. *Gender, Place and Culture*, 21(4), pp. 489–507.

Bain, A., Payne, W. and Isen, J. (2014). Rendering a neighbourhood queer. *Social and Cultural Geography*, 16(4), pp. 424–443.

Baker, R. (1994). *Drag: A history of female impersonation in the performing arts.* New York: New York University Press.

Baldwin, A. (2014). What each of Facebook's 51 new gender options means. [Online]. Available at: www.thedailybeast.com/articles/2014/02/15/the-complete-glossary-of-facebook-s-51-gender-options.html [Accessed 28 November 2016].

Baxter, R. and Brickell, K. (2014). For home unmaking. *Home Cultures*, 11(2), pp. 133–143.

Baydar, G. (2012). Sexualised productions of space. *Gender, Place and Culture*, 19(6), pp. 699–706.

Beale, P. (ed.) (1989). *A concise dictionary of slang and unconventional English.* New York: Macmillian.

Bedwell, A. J. (2016). No Pride in Prisons says no to pride parade. *gayexpress* [Online]. Available at: www.gayexpress.co.nz/2016/02/no-pride-prisons-says-no-pride-parade [Accessed 1 July 2016].

Bell, A. (2014). *Relating indigenous and settler identities: Beyond domination.* Basingstoke: Palgrave MacMillan.

Bell, D. and Binnie, J. (2002). *The sexual citizen: Queer politics and beyond.* Cambridge: Polity Press.

Bell, D. and Valentine, G. (eds) (1995). *Mapping desire: Geographies of sexualities.* London: Routledge.

Bender-Baird, K. (2011). *Transgender employment experiences: Gendered perceptions and the law.* New York: State University of New York Press.

Berg, L. and Kearns, R. (1996). Naming as norming: 'Race', gender and the identity politics in Aotearoa/New Zealand. *Environmental Planning D: Society and Space*, 14(1), pp. 99–122.

Bérubé, A. (1990). *Coming out under fire: The history of gay men and women in World War II.* New York: Free Press.

Besnier, N. (1994). Polynesian gender liminality through time and space. In: G. Herdt (ed.), *Third sex, third gender: Beyond sexual dimorphism in culture and history.* Harlow: Zone Books, pp. 285–328.

Besnier, N. (1997). Sluts and superwomen: The politics of gender liminality in urban Tonga. *Ethnos: Journal of Anthropology*, 62(1–2), pp. 5–31.

Besnier, N. (2002). Transgenderism, locality, and the Miss Galaxy beauty pageant in Tonga. *American Ethnologist*, 29(3), pp. 534–566.

Besnier, N. (2004). The social production of abjection: Desire and silencing among transgender Tongans. *Social Anthropology*, 12(3), pp. 301–23.

Besnier, N. (2011). *On the edge of the global.* California: Stanford University Press.

Besnier, N. and Alexeyeff, K. (eds) (2014). *Gender on the edge: Transgender, gay and other Pacific Islanders.* Honolulu: University of Hawai'i Press.

Bettcher, T. M. (2014). Transphobia. *TSQ Transgender Studies Quarterly,* 1(1), pp. 249–251.

Beyer, G. and Casey, C. (1999). *Change for the better: The story of Georgina Beyer as told to Cathy Casey.* Auckland: Random House.

Binnie, J. (2000). Cosmopolitanism and the sexed city. In: D. Bell, and A. Haddour (eds), *City visions.* Harlow: Prentice Hall, pp. 166–178.

Binnie, J. (2004). *The globalization of sexuality.* London: Sage.

Binnie, J. and Klesse, C. (2011). 'Because it was a bit like going to an adventure park': The politics of hospitality in transnational lesbian, gay, bisexual, transgender and queer activist networks. *Tourist Studies,* 11(2), pp. 157–174.

Bird, S. (2002). Case note: Re Kevin (validity of marriage of transsexual) [2001] FamCA 1074. *Southern Cross Law Review,* 6, pp. 364–371.

Blidon, M. (2009). La Gay Pride entre subversion et banalisation. *Espaces, Populations, Societes,* 2, pp. 305–318.

Blunt, A. and Dowling, R. (2006). *Home.* London and New York: Routledge.

Bockting, W. and Coleman, E. (1992). *Gender dysphoria: Interdisciplinary approaches in clinical management.* New York: The Haworth Press.

Bondi, L. (1992). Gender and dichotomy. *Progress in Human Geography,* 14(3), pp. 438–445.

Bondi, L. (2014). Feeling insecure: A personal account in a psychoanalytic voice. *Social and Cultural Geography,* 15(3), pp. 332–350.

Boris, E. (1998). 'You wouldn't want one of them dancing with your wife': Racialized bodies on the job in WWII. *American Quarterly,* 50(1), pp. 77–108.

Bornstein, K. (1994). *Gender outlaw: On men, women and the rest of us.* New York: Routledge.

Brady, A. (2012). The transgendered Kiwi: Homosocial desire and 'New Zealand identity'. *Sexualities,* 15(3–4), pp. 355–372.

Brickell, K. (2012). 'Mapping' and 'doing' critical geographies of home. *Progress in Human Geography,* 36(2), pp. 575–588.

Brown, M. (2012). Gender and sexuality I: Intersectional anxieties. *Progress in Human Geography,* 36(4), pp. 541–550.

Brown, M. (2014). Gender and sexuality II: There goes the gayborhood? *Progress in Human Geography,* 38(3), pp. 457–465.

Browne, K. (2004). Genderism and the bathroom problem: (Re)materialising sexed sites, (re)creating sexed bodies. *Gender, Place and Culture: A Journal of Feminist Geography,* 11, pp. 331–346.

Browne, K. (2006). 'A right geezer-bird' (man-woman): The sites and sights of 'female' embodiment. *ACME: An International E-Journal for Critical Geographies,* 5(2), pp. 121–143.

Browne, K. (2007a). A party with politics? (Re)making LGBTQ pride spaces in Dublin and Brighton. *Social and Cultural Geography,* 8(1), pp. 63–87.

Browne, K. (2007b). Drag queens and drab dykes: Deploying and deploring femininities. In: K. Browne, L. Lim, and G. Brown, (eds)., *Geographies of sexualities.* Hampshire: Ashgate, pp. 113–124.

Browne, K. (2009). Womyn's separatist spaces: Rethinking spaces of difference and exclusion. *Transactions of the Institute of British Geographers,* 35, pp. 541–556.

Browne, K. (2011). Beyond rural idylls: Imperfect lesbian utopias at Michigan Womyn's music festival. *Journal of Rural Studies*, 27(1), pp. 13–23.

Browne, K. and Bakshi, L. (2013). *Ordinary in Brighton?: LGBT, activisms and the city*. Aldershot: Ashgate.

Browne, K. and Lim, J. (2010). Trans lives in the 'gay capital of the UK'. *Gender, Place and Culture: A Journal of Feminist Geography*, 17(5), pp. 615–633.

Browne, K., Lim, J. and Brown, G. (eds) (2007). *Geographies of sexualities: Theories, practices and politics*. Aldershot: Ashgate.

Browne, K., Nash, C. and Hines, S. (2010). Introduction: Towards trans geographies. *Gender, Place and Culture: A Journal of Feminist Geography*, 17(5), pp. 573–577.

Browne, K., Norcup, J., Robson, E. and Sharp, J. (2013). What's in a name? Removing women from the Women and Geography Study Group. *Area*, 45(1), pp. 7–8.

Brun, C. and Lund, R. (2008). Making a home during crisis: Post-tsunami recovery in a context of war, Sri Lanka. *Singapore Journal of Tropical Geography*, 29(3), pp. 274–87.

Bryne, J. (2014). *License to be yourself*. New York: The Open Society Foundations. [Online]. Available at: www.opensocietyfoundations.org/reports/license-be-yourself [Accessed 1 March 2016].

Butler, J. (1990). *Gender trouble: Feminism and subversion of identity*. New York: Routledge.

Butler, J. (1993). *Bodies that matter: On the discursive limits of 'sex'*. New York: Routledge.

Butler, J. (2001). Doing justice to someone: Sex reassignment and allegories of transsexuality. *GLQ: A Journal of Lesbian and Gay Studies*, 7(4), pp. 621–636.

Butler, J. (2004a). *Undoing gender*. New York: Routledge.

Butler, J. (2004b). *Precarious life: The powers of mourning and violence*. London and New York: Verso.

Butler, J. (2009a). Performativity, precarity and sexual politics. *AIBR. Revista de Antropologia Iberoamericana*, 4(3), pp. I – xiii.

Butler, J. (2009b). *Frames of war: When is life grievable?* London: Verso.

Butler, J. (2011). For and against precarity. *Tidal: Occupy Theory, Occupy Strategy*, 1, pp. 12–13.

Caluya, G. (2008). 'The rice steamer': Race, desire and affect in Sydney's gay scene. *Australian Geographer*, 39(3), pp. 283–292.

Camilleri, R. (2012). Does gay mean white? *The Huffington Post*. [Online]. Available at: www.huffingtonpost.com/2012/07/09/latest-conversations-does_n_1659943.html [Accessed 20 November 2012].

Casey, M. (2007). The queer unwanted and their undesirable 'otherness'. In: K. Browne, J. Lim, and G. Brown (eds), *Geographies of sexualities*. Hampshire: Ashgate, pp. 125–136.

Cava, P. (2016). Cisgender and cissexual. In: N. Naples, N. R. C. Hoogland, M. Wickramasinghe and W. C. A. Wong (eds), *The Wiley Blackwell encyclopedia of gender and sexuality studies*. Hoboken, NJ: John Wiley, pp. 1–4.

Cavanagh, S. (2010). *Queering bathrooms: Gender, sexuality, and the hygienic imagination*. Canada: University of Toronto Press.

Chambers, D. (2008). A postcolonial interrogation of attitudes toward homosexuality and gay tourism: The case of Jamaica. In: M. Daye, D. Chambers, and S. Roberts (eds), *New perspectives in Caribbean tourism*. London: Routledge, pp. 94–114.

Chase, C. (1998). Hermaphrodites with attitude: Mapping the emergence of intersex political activism. *GLQ: A Journal of Lesbian and Gay Studies*, 4(2), pp. 189–211.

Chatterton, P. and Hollands, R. (2003). *Urban nightscapes: Youth cultures, pleasure spaces and corporate power.* London and New York: Routledge.

Chrisler, J. and McCreary, D. (eds) (2010). *Handbook of gender research in psychology.* London: Springer.

Christmas, G. (2010). Research note: Intersexuality, feminism and the case for gender binaries. *Women's Studies Journal*, 24(1), pp. 60–65.

Choi, Y. (2013). The meaning of home for transgendered people. In: Y. Taylor, and M. Addison (eds), *Queer presences and absences: Genders and sexualities in the social sciences.* London: Palgrave Macmillan, pp. 118–140.

Cloke, P. and Johnston, R. (eds) (2005). *Spaces of geographical thought: Deconstructing human geography's binaries.* London: Sage.

Colls, R. and Evans, B. (2014). Making space for fat bodies? A critical account of 'the obesogenic environment'. *Progress in Human Geography*, 38(6), pp. 733–753.

Cooke, H. (2018). LGBTI people will still be invisible on next NZ census. *Stuff.co.nz* [Online]. Available at: www.stuff.co.nz/national/politics/100456579/lgbti-people-will-still-be-invisible-on-next-nz-census [Accessed 12 January 2018].

Cotterill, L. (2013). 'You just keep walking into the pen to get your next sheep ...' An exploration of sheep shearer's experiences and responses to heat in the sheep shearing industry. Masters in Public Health thesis, Dunedin Aotearoa: University of Otago. [Online]. Available at: https://otago.ourarchive.ac.nz/bitstream/handle/10523/4129/CotterillLucyE2013MPH.pdf?sequence=1&isAllowed=y [Accessed 18 October 2017].

Cotton, T. (ed.), (2012). *Transgender migrations: The bodies, borders and politics of transition.* New York: Routledge.

Cooper, A., Law, R., Malthus, R. and Wood, P. (2000). Rooms of their own: Public toilets and gendered citizens in a New Zealand city, 1860–1940. *Gender, Place and Culture: A Journal of Feminist Geography*, 7, pp. 417–433.

Cream, J. (1995). Re-solving riddles: The sexed body. In: D. Bell, and G. Valentine (eds), *Mapping desire: Geographies of sexualities.* London: Routledge, pp. 31–40.

Cromwell, J. (1999). *Transmen and FTMs: Identities, bodies, genders and sexualities.* Urbana: University of Illinois Press.

Currah, P. and Minter, S. (2000). Unprincipled exclusions: The struggle to achieve judicial and legislative equality for transgendered people. *William and Mary Journal of Women and Law*, 7(1), pp. 37–64.

Currah, P., Green, J. and Stryker, S. (2009). *The state of transgender rights in the United States of America.* San Francisco, CA: National Sexuality Resource Center.

D'Ooge, C. (2008). Queer Katrina: Gender and sexual orientation matters in the aftermath of the disaster. [Online]. In B. Willinger (ed.), *Katrina and the women of New Orleans.* New Orleans, LA: Tulane University. Available at: http://tulane.edu/nccrow/publications.cfm, pp. 22–24 [Accessed 20 March 2015].

Dahl, U. (2013). White gloves, feminist fists: Race, nation and the feeling of 'vintage' in femme movements. *Gender, Place and Culture*, 21(5), pp. 604–621.

Datta, A. (2012). *The illegal City: Space, law and gender in a Delhi squatter settlement.* London: Ashgate.

Davidson, M. (2007). Seeking refuge under the umbrella: Inclusion, exclusion, and organizing within the category transgender. *Sexuality Research and Social Policy: Journal of NSRC*, 4(4), pp. 60–80.

Davidson, E. (2014). Responsible girls: The spatialized politics of feminine success and aspiration in a divided Silicon Valley, USA. *Gender, Place and Culture*, 22(3), pp. 390–404.

Davis, K. (ed.) (1997). *Embodied practices: Feminist perspectives on the body*. London: Sage.

Davis, A Y. (2011). *Are prisons obsolete?* New York: Seven Stories Press.

Davy, Z. (2011). *Recognizing transsexuals: Personal, political and medicolegal embodiment*. Farnham: Ashgate.

de Jong, A. (2015). Dykes on bikes: Mobility, belonging and the visceral. *Australian Geographer*, 46(1), pp. 1–13.

Department of Corrections (2017). M.03.05 Transgender and intersex prisoner. [Online]. Available at: www.corrections.govt.nz/resources/policy_and_legislation/Prison-Operations-Manual/Movement/M.03-Specified-gender-and-age-movements/M.03-4.html [Accessed 25 August 2017].

Doan, P. (2007). Queers in the American city: Transgendered perceptions of urban spaces. *Gender, Place, and Culture: A Journal of Feminist Geography*, 14, pp. 57–74.

Doan, P. (2009). Safety and urban environments: Transgendered experiences of the city. *Women and Environment International*, 78/79, pp. 22–25.

Doan, P. (2010). The tyranny of gendered spaces: Living beyond the gender dichotomy. *Gender, Place and Culture: A Journal of Feminist Geography*, 17, pp. 635–654.

Doan, P. (ed.), (2011). *Queerying planning: Challenging heteronormative assumptions and reframing planning practice*. Farnham: Ashgate.

Doan, P. (ed.), (2015). *Planning and LBGTQ communities: The need for inclusive queer spaces*. New York: Routledge.

Doan, P. (2016). To count or not to count: Queering measurement and the transgender community. *Women's Studies Quarterly*, 44(3), pp. 89–110.

Dominey-Howes, D., Gorman-Murray, A. and McKinnon, S. (2013). Queering disasters: On the need to account for LGBTI experiences in natural disaster contexts. *Gender, Place and Culture*, 21(7), pp. 905–918.

Douglas, M. (1980). *Purity and danger: An analysis of the concepts of pollution and taboo*, London: Routledge and Kegan Paul.

Dreger, A. (2004). Shifting the paradigm of intersex treatment. Intersex Society of North America website. [Online]. Available at: www.isna.org/drupal/compare [Accessed March 2014].

Duffy, M., Waitt, G. and Gibson, C. (2007). Get into the groove: The role of sound in generating a sense of belonging at street parades. *Altitude*, 8. [Online]. Available at: www.api-network.com/altitude/article.php?issue=8&nid=8&theme=Eight&in progress [Accessed 20 March 2012].

Duggan, L. (2002). The new homonormativity: The sexual politics of neoliberalism. In: R. Castronovo, R. and D. D. Nelson (eds), *Materializing democracy: Toward a revitalized cultural politics*. Durham: Duke University Press, pp. 175–194.

Duncan, N. (ed.) (1996). *Bodyspace: Destablizing geographies of gender and sexuality*. London: Routledge.

Duncan, J. S. and Lambert, D. (2004). Landscapes of home. In: J. Duncan, N. Johnson, and R. Schein, (eds), *A companion to cultural geography*. Oxford: Blackwell Publishing, pp. 382–404.

Dupuis, A. and Thorns, D. (1998). Home, home ownership and the search for ontological security. *The Sociological Review*, 46(1), pp. 24–47.

Eder, J. (2016). Transgender student Stefani Muollo-Gray: I was told to use the boys' toilet. *Stuff.co.nz* [Online]. Available at: www.stuff.co.nz/national/81079657/Transgender-student-Stefani-Muollo-Gray-I-was-told-to-use-the-boys-toilets [Accessed 31 October 2016].

Edwards, J. (2016). Wellington High, Onslow College get gender-neutral bathrooms. *Stuff.co.nz* [Online]. Available at: www.stuff.co.nz/dominion-post/news/wellington/77457959/Wellington-High-Onslow-College-get-gender-neutral-bathrooms [Accessed 10 October 2016].

Ekins, R. and King, D. (1996). *Blending genders: Social aspects of cross-dressing and sex-changing.* London and New York: Routledge.

Ekins, R. (1997). *Male femaling: A grounded theory approach to cross-dressing and sex-changing.* London and New York: Routledge.

Elliott, S. (1994). *The adventures of Priscilla Queen of the Desert.* PolyGram Filmed Entertainment.

Emerton, R. (2006). Finding a voice, fighting for rights: The emergence of the transgender movement in Hong Kong. *Inter-Asia Cultural Studies*, 7(2), pp. 243–269.

Engle, K. (2013). Carmen Rupe, drag performer, brother keepers, and LGBT activist. AWA: All the kick-ass women the history books left out [Online]. Available at: www.amazingwomeninhistory.com/carmen-rupe/ [Accessed 24 August 2017].

Enke, A. (2012). The education of little cis: Cisgender and the discipline of opposing bodies. In: A. F. Enke (ed.) *Transfeminist perspectives in and beyond transgender and gender studies.* Philadelphia: Temple University Press, pp. 60–77.

Enoka, M. (2016). Calling for gender neutral toilets in schools. *The Wireless* [Online]. Available at: http://thewireless.co.nz/articles/call-for-gender-neutral-toilets-in-schools [Accessed 24 January 2017].

Ettlinger, N. (2007). Precarity unbound. *Alternatives*, 32, pp. 319–340.

European Union (2016). European external action service (EEAS) statement by the spokesperson on LGBTI rights in the United States. [Online]. Available at: https://eeas.europa.eu/headquarters/headquarters-homepage/2810_en [Accessed 4 June 2016].

Farran, S. (2014). Outwith the law in Samoa and Tonga. In: N. Besnier and K. Alexeyeff (eds), *Gender on the edge: Transgender, gay, and other Pacific Islanders.* Honolulu: University of Hawai'i Press, pp. 347–370.

Fausto-Sterling, A. (2000). *Sexing the body: Gender politics and the construction of sexuality.* New York: Basic Books.

Feder, E. K. (2011). Tilting the ethical lens: Shame, disgust, and the body in question. *Hypatia*, 26(3), pp. 632–650.

Ferreri, M., Dawson, G. and Vasudevan, A. (2017). Living precariously: Property guardianship and the flexible city. *Transactions of the Institute of British Institute*, 42(2), pp. 246–259.

Feinberg, L. (1992). *Transgender liberation: A movement whose time has come.* New York: World View Forum.

Feinberg, L. (1993). *Stone butch blues.* Boston: Alyson Books.

Feinberg, L. (1997). *Transgender warriors: Marking history from Jon of Arc to Rupaul.* Boston: Beacon.

Ferreday, D. (2008). Showing the girl: The new burlesque. *Feminist Theory*, 9(1), pp. 47–65.

Foucault, M. (1978). *The history of sexuality: An introduction: Volume I.* New York: Vintage.

Garber, M. (1992). *Vested interests: Cross-dressing and cultural anxiety.* New York: Routledge.

Gatens, M. (1988). Towards a feminist philosophy of the body. In: B. Caine, E. Grosz, and M. de Lepervanches (eds), *Crossing boundaries: Feminisms and critiques of knowledges.* Sydney: Allen and Unwin, pp. 59–70.

Gershenson, O. and Penner, B. (2009). *Ladies and gents: Public toilets and gender.* Philadelphia: Temple University Press.

Gibson, J. (2010). Sexless in the city: A gender revolution. *Sydney Morning Herald*, 12 March 2010. [Online]. Available at: www.smh.com.au/nsw/sexless-in-the-city-a-gender-revolution-20100311-q1l2.html [Accessed 13 March 2010].

Gilmore, R. (2007). *Golden gulag: Prisons, surplus, crisis, and opposition in globalizing California.* Berkeley: University of California Press.

Glamilton Burlesque Academy (2016). NZ Grand Dame of Burlesque 80th Birthday Show at The Meteor 27 May 2016. [Online]. Available at: www.scoop.co.nz/stories/CU1605/S00345/nz-grand-dame-of-burlesque-80th-birthday-show-at-the-meteor.htm [Accessed 8 November 2017].

Glee (2009). Television series. Created by Ian Brennan, Brad Falchuk and Ryan Murphy.

Goodrich, K. M. (2012). Lived experiences of college-age transsexual individuals. *Journal of College Counseling*, 15, pp. 215–232.

Gordon, M., Price, M. and Peralta, K. (2016). Understanding HB2: North Carolina's newest law solidifies state's role in defining discrimination. *The Charlotte Observer* [Online] Available at: www.charlotteobserver.com/news/politics-government/article68401147.html [Accessed 2 April 2016].

Gorman-Murray, A. (2006). Gay and lesbian couples at home: Identity work in domestic space. *Home Cultures*, 3(2), pp. 145–167.

Gorman-Murray, A. (2008). Reconciling self: Gay men and lesbians using domestic materiality for identity management. *Social and Cultural Geography*, 9(3), pp. 283–301.

Gorman-Murray, A. (2012). Queer politics at home: Gay men's management of the public/private boundary. *New Zealand Geographer*, 68(2), pp. 111–120.

Gorman-Murray, A. Johnston, L. and Waitt, G. (2010). Queer(ing) communication in research relationships: A conversation about subjectivities, methodologies and ethics. In: C. J. Nash and K. Browne (eds), *Queer methods and methodologies: Intersecting queer theories and social science research.* Farnham and Burlington: Ashgate, pp. 97–112.

Gorman-Murray, A. McKinnon, S. and Dominey-Howes, D. (2014). Queer domicide: LGBT displacement and home loss in natural disaster impact, response and recovery. *Home Cultures*, 11(2), pp. 237–162.

Grant, J. M., Mottet, L. A. and Tanis, J. (2011) *Injustice at every turn: A report of the national transgender discrimination survey.* Washington, DC: National Center for Transgender Equality and National Gay and Lesbian Task Force.

Grewal, I. and Kaplan. C. (2001). Global Identities: Theorizing transnational studies of Sexuality. *GLQ*, 6(4), pp. 663–79.

Greed, C. (2003). *Inclusive urban design: Public toilets.* Woburn: Architectural Press.

Greenstone Pictures (2002). *Yellow for hermaphrodite – Mani's story.* Producer and director John Keir. Auckland.

Grosz, E. (1994). *Volatile bodies: Towards a corporeal feminism.* Bloomington and Indianapolis: Indiana University Press.

Gruber, R. (2016). 30 Gender-reveal cakes to inspire your big unveiling. [Online]. Available at: www.popsugar.com/moms/Gender-Reveal-Party-Cakes-8400213 #photo-8400213 [Accessed 28 November 2016].

Halberstam, J. (1998). *Female masculinity.* London: Duke University Press.

Halberstam, J. (2005). *In a queer time and place: Transgender bodies, subcultural lives.* New York: New York University Press.

Halberstam, J. (2012). On pronouns. *Jack Halberstam: Gaga feminism and queer failure* [Online]. Available at: www.jackhalberstam.com/on-pronouns [Accessed 20 November 2012].

Haritaworn, J. and Snorton. R. (2013). Transsexual necropolitics. In: A. Aizura and S. Stryker (eds), *The Transgender Studies Reader 2.* New York: Routledge, pp. 66–76.

Harris, J. (2016). *Hui Takatāpui: Our place to stand 30 years on.* New Zealand AIDS Foundation [Online]. Available at: www.nzaf.org.nz/news-and-media/news/hui-takataapui-our-place-to-stand-30-years-on/ [Accessed 18 March 2017].

Hartal, G. (2015). The gendered politics of absence: Homonationalism and gendered power relations in Tel Aviv's gay-centre. In: K. Brown and E. Ferreira (eds), *Lesbian geographies: Gender, place and power.* Farnham: Ashgate, pp. 91–112.

He, C. (2013). Performance and the politics of gender: Transgender performance in contemporary Chinese films. *Gender, Place and Culture*, 21(5), pp. 622–636.

Henig, R. M. (2017). Rethinking gender. *National Geographic* Special Issue 'The Shifting Landscape of Gender', 231(1), pp. 48–72.

Herman, E. (2015). Tranarchism: Transgender embodiment and destabilization of the state. *Contemporary Justice Review*, 18(1), pp. 76–92.

Hetherington, K. (2003). Spatial textures: place, touch and praesentia. *Environment and Planning A*, 35, pp. 1933–1944.

Hines, S. (2007). *Transforming gender: Transgender practices of identity, intimacy and care.* Bristol: The Policy Press.

Hines, S. (2010). Queerly situated? Exploring negotiations of trans queer subjectivities at work and within community spaces in the UK. *Gender, Place and Culture: A Journal of Feminist Geography*, 17(5), pp. 597–613.

Hines, S. (2013). *Gender diversity, recognition and citizenship: Towards a politics of difference.* London: Palgrave McMillian.

Hines, S. (2017). The feminist frontier: On trans and feminism. *Journal of Gender Studies.* DOI: 10.1080/09589236.2017.1411791.

Hines, S. and Sanger, T. (eds) (2010). *Transgender identities: Towards a social analysis of gender diversity.* London and New York: Routledge.

Hines, S., Taylor, Y. and Casey, M. (eds) (2010). *Theorizing intersectionality and sexuality,* London: Palgrave MacMillan.

Hines, S. and Taylor, Y. (2012). *Sexualities: reflections and futures.* London: Palgrave Macmillan.

Hird, M. (2003). Considerations for a psychoanalytic theory of gender identity and sexual desire: The case of intersex'. *Signs*, 28(4), pp. 1067–1092.

Hopkins, P. (2017). Social geography I: Intersectionality. *Progress in Human Geography*, pp. 1–11. DOI: 10.1177/0309132517743677.

Hubbard, P. (2012). *Sexualities and cities.* London: Routledge.

Hutchings, J. and Aspin, C. (2007). *Sexuality and the stories of indigenous people.* Wellington: Huia Publishers.

I am Cait (2015). TV mini-series documentary. E! Channel (2015–2016). Executive producer Andrea Metz.

Ingraham, C. (1992). Initial properties: Architecture and the space of the line. In: B. Colomina (ed.), *Sexuality and space.* New York: Princeton University Press, pp. 255–271.

International Civil Aviation Organisation (2014) Machine readable travel documents, sixth edition, Part 1, Section IV, p. 11. Available at: www.icao.int/Meetings/TAG-MRTD/TagMrtd22/TAG-MRTD-22_WP03-rev.pdf [Accessed 2 December 2016].

International Commission of Jurists (ICJ) (2007). *Yogyakarta principles – principles on the application of international human rights law in relation to sexual orientation and gender identity.* [Online]. Available at: www.icj.org/wp-content/uploads/2012/08/Yogyakarta-Principles-publication-2007-eng.pdf [Accessed 5 January 2016].

Irazábal, C. and Huerta, C. (2016). Intersectionality and planning at the margins: LGBTQ youth of color in New York. *Gender, Place and Culture: A Journal of Feminist Geography*, 23(5), pp. 714–732.

Irigaray, L. (1984). *L'ethique de la différence sexuelle.* Paris: Minuit.

Isoke, Z. (2013). Can't I be seen? Can't I be heard? Black women queering politics in Newark. *Gender, Place and Culture*, 21(3), pp. 353–369.

Jackson, P. (2004). *Inside clubbing: Sensual experiments in the art of being human.* Oxford and New York: Berg.

Jeffreys, S. (2014). *Gender hurts: A feminist analysis of the politics of transgenderism.* London: Routledge.

Johnson, L. (2008). Re-placing gender? Reflections on 15 years of gender, place and culture. *Gender, Place and Culture*, 15(6), pp. 561–574.

Johnston, L. (1995). The politics of the pump: Hard core gyms and women body builders. *New Zealand Geographer*, 51(1), pp. 16–18.

Johnston, L. (1996). Pumped up politics: Female body builders refiguring the body. *Gender, Place and Culture: A Journal of Feminist Geography*, 3(3), pp. 327–340.

Johnston, L. (1997). Queen(s') street or Ponsonby poofters? The embodied HERO parade sites. *New Zealand Geographer*, 53(2), pp. 29–33.

Johnston, L. (1998). Reading sexed bodies in sexed spaces. In: H. Nast, and S. Pile (eds), *Places through the body.* London: Routledge, pp. 244–262.

Johnston, L. (2001). (Other) bodies and tourism studies. *Annals of Tourism Research*, 28(1), pp. 180–201.

Johnston, L. (2005a). Man: Woman. In: P. Cloke, and R. Johnston (eds), *Spaces of geographical thought: Deconstructing human geography's binaries.* London: Sage, pp. 119–141.

Johnston, L. (2005b). *Queering tourism: Paradoxical performances at gay pride parades.* London and New York: Routledge.

Johnston, L. (2006). 'I do down-under': Wedding tourism in Aotearoa New Zealand. *Acme: An International E-Journal for Critical Geographies*, 5(2), pp. 191–208.

Johnston, L. (2009). Scholar's choice essay: *The Topp Twins: Untouchable Girls: The Movie. Emotion, Space and Society*, 2, pp. 70–72.

Johnston, L. (2012). Sites of excess: The spatial politics of touch for drag queens in Aotearoa New Zealand. *Emotion, Space and Society*, 5, pp. 1–9.

Johnston, L. (2014). Abstract: Documenting bodies beyond binaries. *The Association of American Geographers 106th Annual Meeting Abstract Volume*, April 8–12, Tampa, Florida, published on CD. [Online]. Available at: http://meridian.aag.org/callforpapers/program/AbstractDetail.cfm?AbstractID=55359 [Accessed 12 June 2014].

Johnston, L. (2016). Gender and sexuality I: Genderqueer geographies? *Progress in Human Geography*, 40(5), pp. 668–678.

Johnston, L. (2017a). Gender and sexuality II: Activism. *Progress in Human Geography*, 41(5), pp. 648–656.

Johnston, L. (2017b). Gender and sexuality III: Precarious places. *Progress in Human Geography*, pp. 1–9. DOI: 10.1177/0309132517731256.

Johnston, L. and Longhurst, R. (2010). *Space, place and sex: Geographies of sexualities*. Lanham MD: Rowman and Littlefield.

Johnston, L. and Longhurst, R. (2013). Geografias trans(icionais): Corpos, binarismos, lugares e espaços. In: J. M. Silva, M. Ornat and A. Baptista Chimin Junior (eds), *Geografias malditas: Corpos, sexualiades e sspaços*, Todapalavre Editora: Paraná, pp. 339–355.

Johnston, L. and Longhurst, R. (2016). Trans(itional) geographies: Bodies, binaries, places and spaces. In: G. Brown, K. Browne (eds), *Research companion of sex and sexualities*. Farnham: Ashgate, pp. 45–54.

Johnston, L. and Valentine, G. (1995). Wherever I lay my girlfriend, that's my home: The performance and surveillance of lesbian identities in domestic environments. In: D. Bell, and G. Valentine (eds), *Mapping desire: Geographies of sexualities*. London: Routledge, pp. 99–113.

Johnston, L. and Waitt, G. (2015). The spatial politics of gay pride parades and festivals: Emotional activism. In: D. Paternotte, and M. Tremblay, (eds)., *Ashgate research companion on lesbian and gay activism*. Farnham: Ashgate, pp. 105–119.

Jordan, J. (2005). *The sex industry in New Zealand: A literature review*. [Online]. Available at: www.justice.govt.nz/assets/Documents/Publications/sex-industry-in-nz.pdf [Accessed 6 October 2017].

Kaika, M. (2004). Interrogating the geographies of the familiar: Domesticating nature and constructing the autonomy of the modern home. *International Journal of Urban and Regional Research*, 28(2), pp. 265–286.

Kaminski, E. and Taylor, V. (2008). 'We're not just lip-synching up here': Music and collective identity in drag performances. In: J. Reger, R. L. Einwohner, D. J. Myers (eds), *Identity work in social movements*. Minneapolis and London: University of Minnesota Press, pp. 47–76.

Kannen, V. (2013). These are not 'regular places': Women and gender studies classrooms as heterotopias. *Gender, Place and Culture*, 21(1), pp. 52–67.

Keogh, B. (2016). College says yes to trans student using female toilets. *New Zealand Herald* [Online]. Available at: http://m.nzherald.co.nz/nz/news/article.cfm?c_id=1&objectid=11667223 [Accessed 26 July 2016].

Kerekere, E. (2017). *Part of the whānau: The emergence of takatāpui Identity. He whāriki takatāpui*. Doctoral thesis. Wellington Aotearoa: Victoria University of Wellington.

Kessler, S. J. (1998). *Lessons from the intersexed*. New Brunswick: Rutgers University Press.

Kirkland, A. (2006). What's at stake in transgender discrimination as sex discrimination? *Signs: Journal of Women in Culture and Society*, 32(1), pp. 83–111.

Kristeva, J. (1982). *Powers of horror: An essay on abjection*. Translated by Leon S. Roudiez, New York: Columbia University Press.

Kulick, D. (1998). *Travesti: Sex, gender and culture among Brazilian transgendered prostitutes*. Chicago: University of Chicago Press.

Larner, W. and Spoonley, P. (1995). Postcolonial politics in Aotearoa/New Zealand. In: D. Stasiulis and N. Yuval-Davis (eds), *Unsettling settler societies: Articulations of gender, race, ethnicity and class.* London: Sage, pp. 39–64.

Laurier, E. and Philo, C. (2006). Possible geographies: A passing encounter in a café. *Area,* 38(4), pp. 353–363.

Lawson, V. (2005). Hopeful geographies: Imagining ethical alternatives. *Singapore Journal of Tropical Geography,* 26, pp. 36–38.

Lewis, C. (2011). *Cindy Lewis.* Podcast about attending the human rights conference as part of the 2011 Asia Pacific Outgames. [Online]. Available at: www.pridenz.com/apog_cindy_lewis.html?page=1 [Accessed 10 November 2015].

Lewis, H., Dwyer, P., Hodkinson, S. and Waite, L. (2016). Hyper-precarious lives: Migrants, work and forced labour in the Global North. *Progress in Human Geography,* 9(5), pp. 580–600.

Liddicoat, J. (2008). *To be who I am – Ka noho au ki toku anō ao: Report of the inquiry into discrimination experienced by transgender people, He pūrongo mō te uiuitanga mō aukatitanga e pāngia ana e ngā tangata whakawhitiira.* Auckland: New Zealand Human Rights Commission.

Lighter, J. E. (ed.) (1994). *Random House historical dictionary of American slang.* New York: Random House.

Lim, J. (2007). Queer critique and the politics of affect. In: K. Browne, J. Lim, G. Brown (eds), *Geographies of sexualities: Theory, practices, and politics.* Aldershot: Ashgate, pp. 53–68.

Lombardi, E., Wilchins, R. A., Prising Esq., D. and Malouf, D. (2002). Gender violence. *Journal of Homosexuality,* 42(1), pp. 89–101.

Longhurst, R. (2001). *Bodies: Exploring fluid boundaries.* London: Routledge.

Longhurst, R. (2003). Introduction: Placing subjectivities, spaces and places. In: K. Anderson, M. Domosh, S. Pile, N. Thrift (eds), *Handbook of cultural geography.* London: Sage, pp. 282–289.

Longhurst, R. (2005). The body. In: D. Atkinson, P. Jackson, D. Sibley and N. Washbourne (eds), *Cultural geography: A critical dictionary of key concepts.* London and New York: IB. Taurus, pp. 93–98.

Longhurst, R. (2008). *Maternities: gender, bodies, space.* London: Routledge.

Longhurst, R. (2014). Queering body size and shape: Performativity, the closet, shame, and orientation. In: C. Pausé, J. Wykes and S. Murray (eds), *Queering fat embodiment.* Farnham: Ashgate, pp. 13–25.

Longhurst, R. and Johnston, L. (2014). Bodies, gender, place and culture: 21 years on. *Gender, Place and Culture,* 21(3), pp. 267–278.

Longhurst, R., Ho, E. and Johnston, L. (2008). Using the body as an instrument of research: kimch'i and pavlova, *Area,* 40(2), pp. 208–217.

Luamanu, J. (2017). Fa'afatama gathering first of its kind in Samoa. *Samoa Observer* 31 October 2017. [Online]. Available at: http://sobserver.ws/en/01_11_2017/local/26109/Fa%C3%A2%E2%82%AC%E2%84%A2afatama-gathering-first-of-its-kind-in-Samoa.htm [Accessed 10 January 2018].

Lykke, N. (2010). *Feminist studies: A guide to intersectional theory, methodology and writing.* New York: Routledge.

MacPherson, L. (2017). National ethnic population projections: 2013 (base) – 2038 (update). *Statistics New Zealand.* [Online]. Available at: file:///C:/Users/lyndaj/Downloads/NationalEthnicPopulationProjections2013-2038HOTP.pdf [Accessed 5 January 2018].

Mageo, J.-M. (1992). Male transvestism and cultural change in Samoa. *American Ethnologist*, 19(3), pp. 443–459.

Malbon, B. (1999). *Clubbing: Dancing, ecstasy and vitality*. London and New York: Routledge.

Malta Gender Identity, Gender Expression and Sex Characteristics Act (2015). Available at: http://justiceservices.gov.mt/DownloadDocument.aspx?app=lom&itemid=12312&l=1 [Accessed 26 February 2017].

Mann, C. (2013). Big hands 'bad news' for beauty business. *Stuff.co.nz* [Online]. Available at: www.stuff.co.nz/national/8588470/Big-hands-bad-news-for-beauty-business [Accessed 25 April 2014].

Manning, B. (2015). Thousands watch Auckland Pride Parade. *New Zealand Herald* [Online]. Available at: www.nzherald.co.nz/lifestyle/news/article.cfm?c_id=6&objectid=11405808 [Accessed 5 July 2016].

Māori Television (2008). *Takatāpui*, episode 30 June 2008. Produced by Front of the Box Productions.

Markwell, K. and Waitt, G. (2009). Festivals, space and sexuality: Gay pride in Australia. *Tourism Geographies: An International Journal of Tourism Space, Place and Environment*, 11(2), pp. 143–68.

Marriage (Definition of Marriage) Amendment Act (2013). Available at: www.legislation.govt.nz/act/public/2013/0020/latest/DLM4505003.html [Accessed 15 November 2014].

McClintock, A. (1995). *Imperial leather: Race, gender and sexuality in the colonial contest*. New York: Routledge.

McCready, L. T. (2004). Some challenges facing queer youth programs in urban high schools: Racial segregation and de-normalizing whiteness. *Journal of Gay and Lesbian Issues in Education*, 1(3), pp. 37–51.

McDermott, E. (2011). The world some have won: sexuality, class and inequality. *Sexualities*, 14(1), pp. 63–78.

McDonald, L. (2017). Gender issues left out of New Zealand's next census. *Stuff.co.nz* [Online]. Available at: www.stuff.co.nz/the-press/news/95692743/Gender-issues-left-out-of-New-Zealands-next-census [Accessed 20 December 2017].

McDowell, L. (2009). *Working bodies: Interactive service economies and workplace identities*. Oxford: Wiley Blackwell.

McDowell, L. and Court, G. (1994). Performing work: Bodily representations in merchant banks. *Environment and Planning D: Society and Space*, 12, pp. 727–750.

Mead, HM. (2016). *Tikanga Māori: Living by Māori values* (Revised Edition). Wellington: Huia Publishers.

Melville, L. (2016). Lesbians making babies: Why research on sperm, space and decisions matter. *Te Kura Kete Aronui: Graduate and Postgraduate E-Journal*, 7, pp. 1–16. [Online]. Available at: www.waikato.ac.nz/__data/assets/pdf_file/0008/317429/Lisa-Melville-Lesbians-Making-Babies.pdf [Accessed 5 January 2018].

Meyerowitz, J. (2002). *How sex changed: A history of transsexuality in the United States*. Cambridge, MA: Harvard University Press.

Millar, K. (2017). Toward a critical politics of precarity. *Sociology Compass* 11: e12483. [Online]. Available at: http: //doi.org/10.1111/soc4.12483 [Accessed 30 May 2017].

Mitchell, M. B. (2015). Mani B Mitchell: Counsellor, clinical supervisors, mediator, change agent, educator. *Mani B Mitchell Blog*. [Online]. Available at: www.manimitchell.com/ [Accessed 1 February 2016].

Misgav, C. (2015). With the current, against the wind: Constructing spatial activism and radical politics in the Tel-Aviv Gay Center. *ACME: An International Journal for Critical Geographies*, 14(4), pp. 1208–1234.

Misgav, C. and Johnston, L. (2014). Dirty dancing: The (non)fluid geographies of a queer nightclub in Tel Aviv. *Social and Cultural Geography*, 15(7), pp. 730–746.

Momoisea, L. (2015). NZ fa'afafine hope to challenge norms through dance. *Radio New Zealand International*. [Online]. Available at: www.radionz.co.nz/international/programmes/datelinepacific/audio/201779690/nz-fa'afafine-hope-to-challenge-norms-through-dance [Accessed 25 November 2017].

Morrison, C.-A. (2012). Hetero(sex)uality and home: Intimacies of space and places of touch. *Emotion, Space and Society*, 5(1), pp. 10–18.

Moss, P. and Dyck, I. (2002). *Women, body, illness: Space and identity in the everyday lives of women with chronic illness.* Lanham: Rowman and Littlefield.

Munro, S. and Warren, L. (2004). Transgendering citizenship. *Sexualities* 7(3), pp. 345–362.

Munt, S. (1998). *Butch and femme.* London: Cassell.

Munt, S. (2001). The butch body. In: R. Holliday, and J. Hassard (eds), *Contested bodies.* London: Routledge, pp. 95–106.

Nadal, K. L., Davidoff, K. C. and Fujii-Doe, W. (2014). Transgender women and the sex work industry: Roots in systemic, institutional, and interpersonal discrimination. *Journal of Trauma and Dissociation*, 15(2), pp. 169–183.

Nally, C. (2009). Grrly hurly burly: Neo-burlesque and the performance of gender. *Textual Practice*, 23(4): pp. 621–643.

Namaste, K. (1996). Genderbashing: Sexuality, gender, and the regulation of public spaces, *Environment and Planning D*, 14, pp. 221–240.

Namaste, K. (2000). *Invisible lives: The erasure of transsexual and transgendered people.* Chicago, IL: Chicago University Press.

Name withheld (2016). Reader report: Trans people need a safe home. *Stuff.co.nz.* [Online]. Available at: www.stuff.co.nz/stuff-nation/assignments/share-your-news-and-views/13810896/Trans-people-need-a-safe-home [Accessed 1 February 2017].

Nash, C. J. (2010a). 'Queer conversations: old-time lesbians, transmen, and the politics of queer research'. In: K. Brown and C. Nash (eds), *Queer methods and methodologies: Intersecting queer theories and social science research.* Burlington, VT: Ashgate, pp. 129–142.

Nash, C. J. (2010b). Trans geographies, embodiment and experience. *Gender, Place and Culture: A Journal of Feminist Geography*, 17(5), pp. 579–595.

Nash, C. J. (2011). Trans experiences in lesbian and queer space. *The Canadian Geographer*, 55(2), pp. 192–207.

Nash, C. J. and Bain, A. (2007). 'Reclaiming raunch?': Spatializing queer identities at a Toronto women's bathhouse event. *Social and Cultural Geography*, 8, pp. 47–62.

Nash, C. J. and Gorman-Murray, A. (2014). LGBT neighbourhoods and 'new mobilities': Towards understanding transformations in sexual and gendered urban landscapes. *International Journal of Urban and Regional Research*, 38(3), pp. 756–772.

Nast, H. (2002). Queer patriarchies, queer racisms, international. *Antipode*, 34, pp. 874–909.

Nast, H. and Pile, S. (1998). *Places through the body.* London: Routledge.

National Center for Transgender Equality. (2014). *Know your rights: Airport security and transgender people*. [Online]. Available at: www.transequality.org/Resources/AirportSecurity_November2013.pdf [Accessed 3 January 2017].

National Geographic (2017). The shifting landscape of gender, Special Issue, January, 231(1), pp. 48–72.

Nestle, J., Howell, C. and Wilchins, R. (2002). *GenderQueer: Voices from beyond the sexual binary*. Los Angeles, CA: Alyson Books.

New Zealand Statistics (2015). *Statistical Standard for Gender Identity*. [Online]. Available at: file:///C:/Users/lyndaj/Downloads/stat-std-gender-identity.pdf [Accessed 12 January 2017].

New Zealand Statistics (2017). *Historical Context*. [Online]. Available at: www.stats.govt.nz/browse_for_stats/Maps_and_geography/Geographic-areas/urban-rural-profile/historical-context.aspx Accessed 26 April 2017 [Accessed 26 July 2017].

Newshub (2018). LGBT community slams Statistics New Zealand's 'gross incompetence'. [Online]. Available at: www.newshub.co.nz/home/new-zealand/2018/01/lgbt-community-slams-statistics-new-zealand-s-gross-incompetence.html [Accessed 13 January 2018].

Nordmarken, S. (2014). Becoming ever more monstrous: Feeling transgender in-betweenness. *Qualitative Inquiry*, 20(1), pp. 37–50.

NZPC (2013). *What impact has the decriminalisation of sex work in New Zealand had on Māori?* Wellington: New Zealand Prostitutes Collective.

Obrador-Pons, P. (2007). A haptic geography of the beach: Naked bodies, vision and touch. *Social and Cultural Geography*, 8(1), pp. 123–141.

Orange is the New Black (2013). Television series, Netflix. Created by Jenji Kohan.

Oswin, N. (2010). The modern model family at home in Singapore: A queer geography. *Transactions of the Institute of British Geographers*, 35, pp. 256–268.

Oxford English Dictionary (1989). Entry on drag [Online]. Available at: http://dictionary.oed.com/cgi/entry/50069468?query_type=word&queryword=drag&first=1&max_to_show=10&sort_type=alpha&result_place=1&search_id=XTau-pT6tUK-6750&hilite=50069468 [Accessed 17 July 2009].

Pacific Media Watch (2016). Samoa: *Sunday Samoan* condemned for 'disgusting, degrading' reporting of death. [Online]. Available at: www.pmc.aut.ac.nz/pacific-media-watch/samoa-sunday-samoa-condemned-disgusting-degrading-reporting-death-9705 [Accessed 22 June 206].

Papa, R. and Meredith, P. (2012). Kīngitanga – the Māori King movement – Origins of the Kīngitanga. *Te Ara – the encyclopedia of New Zealand*. [Online] Available at: www.TeAra.govt.nz/en/kingitanga-the-maori-king-movement/page-1 [Accessed 18 September 2017].

Paterson, M. (2009). Haptic geographies: Ethnography, haptic knowledges and sensuous dispositions. *Progress in Human Geography*, 33(6), pp. 766–788.

Payne, W. (2016). Death-squads contemplating queers as citizens: what Colombian paramilitaries are saying. *Gender, Place and Culture*, 23(3), pp. 28–344.

Pega, F., Reisner, S., Sell, R. and Veale, J. (2017). Transgender health: New Zealand's innovative statistical standard for gender identity. *American Journal of Public Health*, 107(2), pp. 217–221.

Philo, C. (2014). Insecure bodies/selves: Introduction to themed section. *Social and Cultural Geography*, 15(3), pp. 284–290.

Podmore, J. (2013). Lesbians as Village Queers: The Transformation of Montréal's Lesbian Nightlife in the 1990s. *ACME*, 12(2), pp. 220–249.

Podmore, J. (2015). Contested dyke rights to the city: Montréal's 2012 dyke marches in time and space. In: K. Browne, and E. Ferreira (eds), *Lesbian geographies: Gender, power and place.* Farnham: Ashgate, pp. 71–90.

Ponsonby Productions Limited (2012). *Intersexion.* Writer, director, editor Grant Lahood. Producer John Keir, Researcher, presenter Mani Mitchell. See www.intersexionfilm.com/

Porteous, J. D. (1986). Intimate sensing. *Area*, 25(1), pp. 11–16.

Preves, S. E. (2003). *Intersex and identity: The contested self.* New Brunswick: Rutgers University Press.

Probyn, E. (2003). The spatial imperative of subjectivity. In: K. Anderson, M. Domosh, S. Pile and N. Thrift (eds), *Handbook of cultural geography.* London: Sage, pp. 290–299.

Probyn, E. (2005). *Blush: Faces of shame.* Minneapolis, MN: University of Minnesota Press.

Prosser, J. (1998). *Second skins: The body narratives of transsexuality.* New York: Columbia University Press.

Prosser, J. (2006). Judith Butler: Queer feminism, transgender, and the transubstantiation of sex. In: S. Stryker and S. Whittle (eds), *The transgender studies reader.* New York and London: Routledge, pp. 257–280.

Puar, J. (2005). Queer times, queer assemblages. *Social Text*, 23(3–4), pp. 121–139.

Puar, J. (2012). Precarity talk: A virtual roundtable with Lauren Berlant, Judith Butler, Bojana Cvejić, Isabell Lorey, Jasbir Puar, and Anan Vujanović. *TDR: The Drama Review*, 56(4), pp. 163–177.

Radio New Zealand (2016a). Corrections staff to join Pride Parade [Online]. Available at: www.radionz.co.nz/news/regional/296282/corrections-staff-to-join-pride-parade [Assessed 11 February 2016].

Radio New Zealand (2016b). Police and protesters clash at Auckland Parade [Online]. Available at: www.radionz.co.nz/news/regional/297031/police-and-protesters-clash-at-akl-pride [Accessed 20 February 2016].

Radio New Zealand (2017). Same-sex marriage not a priority for Samoa Fa'afafine Association, 13 December 2017. [Online]. Available at: www.radionz.co.nz/international/pacific-news/346065/same-sex-marriage-not-a-priority-for-samoa-fa-afafine-association [Accessed 15 December 2017].

Rākete, E. and NPIP (2015). Auckland pride controversy: No Pride in Prison responds. *Gaynz.com* [Online]. Available at: www.gaynz.com/articles/publish/45/article_16997.php [Accessed 20 June 2016].

Raun, T. (2012). *Out online: Trans self-representation and community building on You Tube.* PhD dissertation. Department of Culture and Identity: Roskilde University. [Online]. Available at: http://nordicom.statsbiblioteket.dk/ncom/files/30336545/Tobias_final_with_front_page_pfd.pdf [Accessed 8 May 2017].

Raymond, J. (1974). *The transsexual empire: The making of the she-male.* Boston, MA: Beacon Press.

Recher, A. (2015). *Malta adopts ground-breaking trans and intersex law – TGEU press release.* [Online]. Available at: http://tgeu.org/malta-adopts-ground-breaking-trans-intersex-law/ [Accessed 28 February 2017].

Reeve, D. (2014). Part of the problem or part of the solution? How far do 'reasonable adjustments' guarantee 'inclusive access for disabled customers'? In: K. Soldatic, H. Morgan and A. Roulstone (eds), *Disability, spaces and places of policy exclusion.* London and New York: Routledge, pp. 99–114.

Reck, J. (2009). Homeless gay and transgender youth of color in San Francisco. *Journal of LGBT Youth*, 6, pp. 223–242.

Refuge Restrooms (2017). [Online] Available at www.refugerestrooms.org/ [Accessed 19 January 2017].

Rich, A. (1986). Notes towards a politics of location. In: *Blood, bread and poetry: Selected prose 1979–1985*. New York: W. W. Norton and Company, pp. 210–231.

Riviere, J. (1986). Womanliness as masquerade. In: V. Burgin, J. Donald and C. Kaplan (eds), *Formations of fantasy*. London: Methuen, pp. 35–44.

Robinson, V. Hockey, J. and Meah, A. (2004). What I used to do ... on my mother's settee: Spatial and emotional aspects of heterosexuality in England. *Gender, Place and Culture: A Journal of Feminist Geography*, 11(2), pp. 417–435.

Rodaway, P. (1994). *Sensuous geographies: Body, sense and place*. London: Routledge.

Rodriguez, D. (2006). *Forced passages: Imprisoned radical intellectuals and the U.S. prison regime*. Minneapolis, M: University of Minnesota Press.

Roen, K. (2001). 'Either / or' and 'both / neither': Discursive tensions in transgender politics. *Signs: Journal of Women in Culture and Society*, 27(2), pp. 501–522.

Roen, K. (2006). Transgender theory and embodiment: The risk of racial marginalization. In: S. Stryker and S. Whittle (eds), *The transgender studies reader*. New York: Routledge, pp. 656–665.

Rose, G. (1993). *Feminism and geography: The limits of geographical knowledge*. Minneapolis, MN: Minnesota Press.

Rose, G. (2004). 'Everyone's cuddled up and it just looks really nice': An emotional geography of some mums and their family photos. *Social and Cultural Geography*, 5(4), pp. 549–564.

Rosenberg, R. and Oswin, N. (2015). Trans embodiment in carceral space: Hypermasculinity and the US prison industrial complex. *Gender, Place and Culture: A Journal of Feminist Geography*, 22(9), pp. 1269–1287.

Rowse, D. (2017). Still no action plan from Corrections. *express*, 20 July 2017. [Online]. Available at www.gayexpress.co.nz/2017/07/still-no-transgender-action-plan-corrections/ [Accessed 25 August 2017].

Rubin, H. (2003). *Self-made men: Identity and embodiment among transsexual men*. Nashville, TN.: Vanderbilt University Press.

Rupp, L. and Taylor, V. (2003). *Drag queens at the 801 Cabaret*. Chicago and London: The University of Chicago Press.

Rushbrook, D. (2002). Cities, queer space and the cosmopolitan tourist. *GLQ: A Journal of Lesbian and Gay Studies*, 8(1–2): pp. 183–206.

Rutledge, L. W. (1999). In praise of old queens. *Celebrate!* August 27, p. 3.

Salmond, J. (1986). *Old New Zealand houses 1800 – 1940*. Auckland: Reed Methuen.

Samoan Fa'afafine Association (2016). To Samoa Observer; we must do better. A statement from the Samoa Fa'afafine Association Inc, 24 June 2016. *Sunday Samoan*. [Online]. Available at: www.samoaobserver.ws/en/24_06_2016/local/7786/'We-need-to-do-better-we-must-do-better'.htm [Accessed 12 July 2016]

Sandell, J. (2010). Transnational ways of seeing: Sexual and national belonging in Hedwig and the Angry Inch. *Gender, Place and Culture*, 17(2), pp. 231–247.

Savea Sano Malifa, G. (2016). Apology to our readers. *Samoa Observer* 21 July. [Online]. Available at: www.sobserver.ws/en/21_06_2016/ local/7674/Apology-to-our-readers.htm [Accessed 22 July 2016].

Schilt, K. (2010). *Just one of the guys? Transgender me and the persistence of gender inequality*. Chicago, IL: University of Chicago Press.

Schilt, K. and Connell, C. (2007). Do workplace gender transitions make gender trouble? *Gender, Work and Organization*, 14(6), pp. 596–618.

Seamon, D. (1980). Body-subject: Time-space routines, and place ballets. In: A. Buttimer, and S. Seamon (eds), *The human experience of space and place*. London: Croon Helm, pp. 148–165.

Senelick, L. (2000). *The changing room: Sex, drugs and theatre.* New York: Routledge.

Shabazz, R. (2009). 'So high you can't get over it, so low you can't get under it': Carceral spatiality and black masculinities in the United States and South Africa. *Souls: A Critical Journal of Black Politics, Culture, and Society*, 11(3), pp. 276–294.

Shah, S. (2015). Queering critiques of neoliberalism in India: Urbanism and inequality in the era of transnational 'LGBTQ' rights. *Antipode*, 47(3), pp. 635–651.

Shapiro, E. (2004). Trans'cending barriers: Transgender organizing on the internet. *Journal of Gay and Lesbian Social Services: Special Issue on Strategies for Gay and Lesbian Rights Organizing*, 16(3–4), pp. 165–179.

Sheehan, R. and Vadjunec, J. (2016) Roller derby's publicness: Toward greater recognition of diverse genders and sexualities in the Bible Belt. *Gender, Place and Culture: A Journal of Feminist Geography*, 23(4), pp. 537–555.

Schmidt, J. (2001). Redefining Fa'afafine: Western discourses and the construction of transgenderism in Samoa. *Intersections: Gender, History and Culture*, (6), pp. 1–16, Available at: intersections.anu.edu.au/issue6/schmidt.html [Accessed 24 February 2014].

Schmidt, J. (2003). Paradise lost? Social change and Fa'afafine in Samoa. *Current Sociology*, 51(3–4), pp. 417 – 432.

Schmidt, J. (2010). *Migrating genders: Westernization, migration, and Samoan Fa'afaine*, Oxford: Ashgate.

Schmidt, J. (2016). Being 'like a woman': Fa'afāfine and Samoan masculinity. *The Asia Pacific Journal of Anthropology*, 17, pp. 287 – 304.

Schmidt, J. (2017). Translating transgender: Using western discourses to understand Samoan fa'afāfine. *Sociology Compass*, 11(5), pp. 1–17. Available at: https://doi.org/10.1111/soc4.12485.

Schoeffel, P. (2014). Representing fa'afāfine: Sex, socialization, and gender identity in Samoa. In: N. Besnier and K. Alexeyeff (eds), *Gender on the edge: Transgender, gay, and other Pacific Islanders*. Honolulu: University of Hawai'i Press, pp. 73–90.

Schroeder, C. G. (2013). (Un)holy Toledo: Intersectionality, interdependence, and neighborhood (trans)formation in Toledo, Ohio. *Annals of the Association of American Geographers*, 104(1), pp. 166–181.

Selen, E. (2012). The stage: A space for queer subjectification in contemporary Turkey. *Gender, Place and Culture*, 19(6), pp. 730–749

Seuffert, N. (2009). Reflections on transgender immigration. *Griffith Law Review*, 18, 2, pp. 428–452.

Sharp, J. (2009). Geography and gender: What belongs to feminist geography? Emotion, power and change. *Progress in Human Geography*, 33(1), pp. 74–80.

Shipp, T., Shipp, D., Bromley, B., Sheahan, R., Cohen, A., Lieberman, E. and Benacerraf, B. (2004). What factors are associated with parents' desire to know the sex of their unborn child? *Birth*, 31(4), pp. 272–279.

Silva, J. M., and Ornat, M. J. (2014). Intersectionality and transnational mobility between Brazil and Spain in travesty prostitution networks. *Gender, Place and Culture*, 22(8), pp. 1–16.

Smith, L., Nairn, K. and Sandretto, S. (2016). Complicating hetero-normative spaces at school formals in New Zealand. *Gender, Place and Culture*, 23(5), pp. 589–606.

Socías, M. E., Marshall, B. D. L., Arístegui, I., Zalazar, V., Romero, M. and Sued, O. (2014). Towards full citizenship: Correlates of engagement with the gender identity law among transwomen in Argentina. *PLoS ONE*, 9(8), pp. 1 – 6. [Online]. Available at: e105402. DOI:10.1371/journal.pone.0105402 [Accessed 20 November 2016].

Somerville, P. (1992). Homelessness and the meaning of home: Rooflessness or rootlessness? *International Journal of Urban and Regional Research*, 16(4), pp. 529–539.

Sontag, S. (2002). Notes on 'camp'. In: F. Cleto (ed.), *Camp: Queer aesthetics and the performing subject*. Ann Arbour: University of Michigan Press, pp. 53–65.

Spade, D. (2003). Resisting medicine, re/modelling gender. *Berkeley Women's Law Journal*, 18(1), pp. 15–37.

Spade, D. (2008). Keynote address: Trans law and politics on a neoliberal landscape. [Online]. Available at: http://zinelibrary.info/files/TransLawPolitics.pdf Accessed 4 September 2016. [Accessed 12 January 2016].

Spade, D. (2011). *Normal life: Administrative violence, critical trans politics, and the limits of the law*. Brooklyn, NY: South End Press.

Springgay, S. (2003). Cloth as intercorporeality: Touch, fantasy and performance and the construction of body knowledge. *International Journal of Education*. [Online]. Available at: www.ijea.org/v4n5/ [Accessed 20 November 2011].

Stallybrass, P. and White, A. (1986). *The politics and poetics of transgression*. London: Methuen.

Stone, S. (1987). *The empire strikes back: A posttranssexual manifestatio*. [Online]. Available at: http://pendientedemigracion.ucm.es/info/rqtr/biblioteca/Transexualidad/trans%20manifesto.pdf [Accessed 3 June 2015).

Stryker, S. (1994). My words to Victor Frankenstein above the village of Chamounix: Performing transgender rage. *GLQ*, 1, pp. 237–254.

Stryker, S. (2006). (De)subjugated knowledges: An introduction to transgendered studies. In S. Stryker and S. Whittle (eds), *The transgender studies reader*. London: Routledge, pp. 1–18.

Stryker, S. (2008). *Transgender history*. Berkeley, CA: Seal Press.

Stryker, S., Currah, P. and Moore, L. G. (2008). Introduction: Trans-, trans, or transgender? *Women's Studies Quarterly*, 36(3/4), pp. 11–22.

Stryker, S. and Aizura, A. (eds) (2013). *The transgender studies reader 2*. New York: Routledge.

Stryker, S. and Bettcher, T. (2016). Introduction: Trans/feminisms. *TSQ Transgender Studies Quarterly*, 3(1–2), pp. 5–14.

Stryker, S. and Currah, P. (2014). Introduction. *TSQ Transgender Studies Quarterly*, 1(1–2), pp. 1–18.

Stryker, S., Paisley, C. and Moore, L. J. (2008). Introduction: Trans-, trans, or transgender? *WSQ: Women's Studies Quarterly*, 36(3–4), pp. 11–22.

Stryker, S. and Whittle, S. (eds) (2006). *The transgender studies reader*. New York: Routledge.

Suckling, L. (2016). Is it OK to judge flatmates on sexual or gender identity? *Stuff.co.nz* [Online]. Available at: www.stuff.co.nz/life.../is-it-okay-to-judge-flatmates-on-sexual-or-gender-identity [Accessed 28 February 2016].

Sudbury, J. (2002). Celling black bodies: Black women in the global prison industrial complex. *Feminist Review*, 70(1), pp. 57–74.

Sudbury, J. (ed.) (2005). *Global lockdown: Race, gender, and the prison-industrial complex*. London: Routledge.

Sultana, F. (2007). Reflexivity, positionality and participatory ethics: Negotiating fieldwork dilemmas in international research. *ACME: An International E-Journal for Critical Geographies*, 6(3), pp. 374–385.

Sylvia Rivera Law Project. (2007). *'It's war in here:' A report on the treatment of transgender and intersex people in New York State's men's prisons*. New York: Sylvia Rivera Law Project.

Szydlowski, M. (2016). Gender recognition and the rights to health and health care: Applying the principle of self-determination to transgender people. *International Journal of Transgenderism*, 17(3–4), pp. 199–211.

Tan, Q. (2013). Flirtatious geographies: Clubs as spaces for the performance of affective heterosexualities. *Gender, Place and Culture*, 20(6), pp. 718–736.

Taylor, V. and Rupp, L. (2005). When the girls are men: Negotiating gender and sexual dynamics in a study of drag queens. *Signs: Journal of Women in Culture and Society*, 30(4), pp. 2115–2139.

Tcherkézoff, S. (2014). Transgender in Samoa: The cultural production of gender inequality. In: N. Besnier, and K. Alexeyeff (eds), *Gender on the edge: Transgender, gay, and other Pacific Islanders*. Honolulu: University of Hawai'i Press, pp. 115–134.

Te Awekotuku, N. (1991). *Mana Wāhine Māori: Selected writings in Māori women's art, culture and politics*. Auckland: New Women's Press.

Te Awekotuku, N. (2001). Hinemoa: Retelling a famous romance. *Journal of Lesbian Studies*. 5(102), pp. 1–11.

Te Rākei Whakaehu (2016). *Te Rākei Whakaehu Official Page*. [Online]. Available at: www.facebook.com/groups/1859212104305978/about/ [Accessed 24 October 2017].

TGEU (2014). *Historic Danish gender recognition law comes into force*. Available at: http://tgeu.org/tgeu-statement-historic-danish-gender-recognition-law-comes-into-force/ [Accessed 14 September 2016].

TGEU (2015). Ireland adopts progressive gender recognition law. Available at: http://tgeu.org/ireland-adopts-progressive-gender-recognition-law/ [Accessed 15 September 2016].

The Dominion Post (2003). Transsexual MP makes last minute appeal to Parliament, 26 June [Online]. Available at: https://walnet.org/csis/news/world_2003/nzpa-030625.html [Accessed 12 June 2011].

The Samoa Observer (2017). Fa'afafine Association new executive, 14 October 2017. [Online]. Available at: www.samoaobserver.ws/en/14_10_2017/local/25423/Fa%E2%80%99afafine-Association-new-Executive.htm [Accessed 9 January 2018].

Thomas, C. (2008). *Masculinity, psychoanalysis, straight queer theory: Essays on abjection in literature*. Urbana: University of Illinois Press.

Thomas, R. (2016). Trade Me users hit back at criticism over 'no heterosexuals' listing for Wellington flat. *Stuff.co.nz*. [Online]. Available at: www.stuff.co.nz/life-style/home-property/76252777/Trade-Me-users-hit-back-at-criticism-over-no-heterosexuals-listing-for-Wellington-flat [Accessed 1 February 2017].

Transgender Law Center (2009). *State of transgender California: Economic health of transgender Californians*. [Online]. Available at: www.transgenderlawcenter.org [Accessed 26 July 2011].

Transparent (2014). An Amazon original series. Created by Jill Soloway.

Tuan, Y.-F. (1974). *Topophilia: A study of environmental perception, attitudes and values.* New Jersey: Eaglewood Cliffs.

Tunåker, C. (2015). 'No place like home': Locating homeless LGBT youth. *Home Cultures* 12(2), pp. 214–259.

Tyner, J. (2016). Population geography III: Precarity, dead peasants, and truncated lives. *Progress in Human Geography*, 40(2), pp. 275–289.

U.S. State Department Foreign Affairs Manual (2016). 7 FAM 300 Appendix M: Gender Change: Available at: https://fam.state.gov/fam/07fam/07fam1300apM.html [Accessed 20 February 2017].

Van Doorn, N. (2013). Architects of 'the good life': Queer assemblages and the composition of intimate citizenship. *Environment and Planning D: Society and Space*, 31, pp. 157–173.

Valentine, D. (2007). *Imagining transgender: An ethnography of a category.* Durham NC: Duke University Press.

Valentine, G. (1997). A safe place to grow up? Parenting perceptions of children's safety and the rural idyll. *Journal of Rural Studies*, 13, pp. 137–148.

Valentine, G. (2001). *Social geographies: Space and society.* Harlow: Pearson Education.

Valentine, G. (2007). Theorizing and researching intersectionality: A challenge for feminist geography. *The Professional Geographer*, 59(1), pp. 10–21.

Valentine, G. (2010). Prejudice: Rethinking geographies of oppression. *Social and Cultural Geography*, 11(6), pp. 519–537.

Vanderbeck, R. M., Andersson, J., Valentine, G., Sadgrove, J. and Ward, K. (2011). Sexuality, activism, and whiteness in the Anglican Communion: The 2008 Lambeth Conference of Anglican Bishops. *Annals of the Association of American Geographers*, 101(3), pp. 670-689.

Vasudevan, A. (2015). The makeshift city: Towards a global geography of squatting. *Progress in Human Geography*, 39(3), pp. 338–259.

Veale, J. (2017). Reflections on transgender representation in academic publishing. *International Journal of Transgenderism*, pp. 1–2. DOI: 10.1080/15532739.2017.1279868.

Vitulli, E. W. (2013). Queering the carceral: Intersecting queer/trans studies and critical prison studies. *GLQ*, 19(1), pp. 111–123.

Waikato-Tainui Games (2016). [Online]. Available at: https://www.facebook.com/events/1507629946231182/ [Accessed13 May 2017].

Waitt, G. (2008). Boundaries of desire: Becoming sexual through the spaces of Sydney's 2002 Gay Games. *Annals of the Association of American Geographers*, 96(4), pp. 773–787.

Waite, L. (2009). A place and space for a critical geography of precarity? *Geography Compass*, 3(1), pp. 412–433.

Warner, M. (1991). Introduction: Fear of a queer planet. *Social Text*, 29, pp. 3–17.

Weaver, H. (2014). Friction in the interstices: Emotion and landscape in *Stone Butch Blues, Emotion, Space and Society*, 12, pp. 85–91.

Weeks, J. (2015). Gay liberation and its legacies. In: D. Paternotte, and M. Tremblay (eds), *Ashgate research companion on lesbian and gay activism.* Farnham: Ashgate, pp. 44–58.

Wetherell, M. (2008). Subjectivity or psycho-discursive practices? Investigating complex intersectional identities. *Subjectivity*, 22, pp. 73–81.

Westbrook, L. and Schilt, K. (2014). Doing gender, determining gender: Transgender people, gender panics, and the maintenance of the sex/gender/sexuality system. *Gender and Society*, 28(1), pp. 32–57.

Whitley, C. (2013). Trans-kin undoing and redoing gender: Negotiating relational identity among friends and family of transgender persons. *Sociological Perspectives*, 56(4), pp. 597–621.

Whittle, S. (2002). *Respect and equality: Transsexual and transgender rights*. London: Cavendish Publishing Company.

Whittle, S., Turner, L. and Al-Alami, M. (2007). *Engendered penalties: Transgender and transsexual people's experiences of inequality and discrimination*. Wetherby: Communities and Local Government Publications.

Wilchins, R. A. (1997a). The GID controversy: Gender identity disorder diagnosis harms transsexuals. *Transgender Tapestry*, 79(31), pp. 44–45.

Wilchins, R. A. (1997b). *Read my lips: Sexual subversion and the end of gender*. Riverdale, NY: Magnus Books.

Wilchins, R. A. (2012). Op-ed: Transgender dinosaurs and the rise of the genderqueers. *The Advocate* [Online]. Available at: www.advocate.com/commentary/riki-wilchins/2012/12/06/transgender-dinosaurs-and-rise-genderqueers [Accessed 4 May 2017].

Wilkinson, E. (2014). Single people's geographies of home: Intimacy and friendship beyond 'the family'. *Environment and Planning A*, 46(10), pp. 2452–2468.

Williams, W. (1971). *A dictionary of the Māori language. Seventh Edition*. First Edition 1844. Second Edition 1852. Third Edition 1871. Fourth Edition 1892. Fifth Edition 1917. Sixth Edition 1957. Wellington: Government Print.

Williams, E. (2013). Sex work and exclusion in the tourist districts of Salvador, Brazil. *Gender, Place and Culture*, 21(4), pp. 453–470.

Williams, C. (2016). Radical inclusion: Recounting the trans inclusive history of radical feminism. *TSQ: Transgender Studies Quarterly* 3(1–2), pp. 254–258.

Wilson, M. (2002). 'I am the Prince of Pain, for I am a princess in the brain': Liminal transgender identities, narratives and the elimination of ambiguities. *Sexualities* 5(4), pp. 425–448.

Woon, C. (2011). Undoing violence, unbounding precarity: Beyond the frames of terror in the Philippines. *Geoforum*, 42, pp. 285–296.

Woon, C. (2014). Precarious geopolitics and the possibilities of nonviolence. *Progress in Human Geography*, 38(5), pp. 654–670.

Wright, L. (1960). *Clean and decent: The fascinating history of the bathroom and the WC*. Toronto: Toronto University Press.

Yanow, D. (2015). *Constructing 'race' and 'ethnicity' in America: Category-making in public policy and administration*. New York: Routledge.

Yeoman, S. (2016). Protesters bring Pride Parade to a halt. New Zealand Herald. [Online]. Available at: www.nzherald.co.nz/nz/news/article.cfm?c_id=1&objectid=11592972 [Accessed 21 February 2016].

Young, I. M. (1990). *Justice and the politics of difference*. Princeton, NJ: Princeton University Press.

Young, I. M. (1997). *Intersecting voices: Dilemmas of gender, political philosophy and policy*. Princeton, NJ: Princeton University Press.

Young, I. M. (2005). *On female body experiences: 'Throwing like a girl' and other essays*. New York: Oxford University Press.

Index

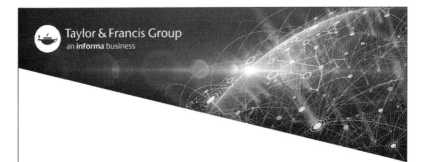

Printed and bound by CPI Group (UK) Ltd, Croydon, CR0 4YY

24/10/2024

01778282-0006